T0179147

SURFACE and COLLOID CHEMISTRY

Principles and Applications

SURFACE and COLLOID CHEMISTRY

Principles and Applications

K. S. Birdi

CRC Press
Taylor & Francis Group
Boca Raton London New York

CRC Press is an imprint of the
Taylor & Francis Group, an **informa** business

CRC Press
Taylor & Francis Group
6000 Broken Sound Parkway NW, Suite 300
Boca Raton, FL 33487-2742

First issued in paperback 2020

ISBN-13: 978-1-4200-9503-6 (hbk)
ISBN-13: 978-0-367-51597-3 (pbk)

This book contains information obtained from authentic and highly regarded sources. Reasonable efforts have been made to publish reliable data and information, but the author and publisher cannot assume responsibility for the validity of all materials or the consequences of their use. The authors and publishers have attempted to trace the copyright holders of all material reproduced in this publication and apologize to copyright holders if permission to publish in this form has not been obtained. If any copyright material has not been acknowledged please write and let us know so we may rectify in any future reprint.

Library of Congress Cataloging-in-Publication Data

Birdi, K. S., 1934-
    Surface and colloid chemistry : principles and applications / K.S. Birdi.
      p. cm.
    Includes bibliographical references and index.
    ISBN 978-1-4200-9503-6 (hardcover : alk. paper)
    1. Surface chemistry. 2. Colloids. I. Title.

QD506.B534 2010
541'.33--dc22
                                          2009028978

Visit the Taylor & Francis Web site at
http://www.taylorandfrancis.com

and the CRC Press Web site at
http://www.crcpress.com

Contents

Preface

The aim of this book is to introduce to the principles and applications of a branch of chemistry called *surface and colloid chemistry*. Most science students are taught physicochemical principles pertaining to *gases, liquids,* and *solids.* The matter around us is recognized to be made of these three states of matter. However, in university chemistry textbooks, seldom more than a chapter is devoted to the science of surface and colloid chemistry. At technical schools worldwide, the same is the case, in general. However, in the realm of applications of physico-chemical technology, the science of *surface and colloid chemistry* is one of the most important. Common examples of the principles at work in this field are:

Rain drops
Combustion engines
Soap bubbles
Foam (in firefighting)
Food products (milk, cheese)
Air pollution (fog, smog, sandstorms)
Wastewater treatment
Washing and cleaning
Cosmetics
Paint and printing; adhesion; friction
Oil and gas production
Oil spill
Plastics and polymers
Biology and pharmaceutical
Milk products
Cement
Adhesive
Coal (coal slurry transport)

Science students are increasingly interested in the application studies of real-world systems. Colloid and surface chemistry offers many opportunities to apply this knowledge in understanding everyday and industrial examples.

The main aim of this book is to guide chemistry or physics students with backgrounds in the area to the level where they are able to understand many natural phenomena and industrial processes, and are able to participate in wider areas of research. The text is carefully arranged so that much of the involved mathematical treatment of this subject is given as references, but it still offers a good understanding of the fundamental principles involved.

Further, this book contains useful data from real-world examples that explain and stimulate the reader to consider both the fundamental theory and industrial applications. The latter is expanding rapidly and every decade brings new application areas in this science. Accordingly, pertinent references are provided for the more advanced student and scientist from other fields (such as biology, geology, pharmacology, medical science, astronomy, plastics, etc.).

Important sample questions and answers are included wherever appropriate in various sections, with detailed data and discussion. Although the text has been primarily aimed at students, researchers will also find some areas of the topics stimulating.

A general high school background in chemistry or physics is all that is required to follow the main theme of this book.

In the past decades, it has become more and more obvious that students and scientists of chemistry and engineering should have some understanding of surface and colloid chemistry. The textbooks on physical chemistry tend to introduce this subject insufficiently. Modern nanotechnology is another area where the role of surface and chemistry is found of much importance. Medical diagnostics applications are also extensive, where both microscale and surface reactions are determined by different aspects of surface and colloid chemical principles. Drug delivery is much based on lipid vesicles (self-assembly structure) that are stabilized by various surface forces.

The book presents some basic considerations regarding liquid surfaces. After an introduction, the liquid–solid interface phenomena is described. Following this, the colloid chemistry systems are discussed, followed by emulsion science and technology. In the last chapter, more complex application examples are described. These are examples where different concepts of surface and colloid chemistry are involved in some mixed manner.

The Author

Professor K. S. Birdi received his BSc (Hons) in chemistry from Delhi University in 1952, and then later traveled to the United States for further studies, majoring in chemistry at the University of California at Berkeley. After graduation in 1957, he went to Standard Oil of California, Richmond. Dr. Birdi moved to Copenhagen, Denmark, in 1959, where he joined the Lever Brothers Development Laboratory in 1959 as chief chemist. During this period, he became interested in surface and colloid chemistry and joined the Institute of Physical Chemistry (founded by Professor J. Brønsted), Danish Technical University, Lyngby, Denmark, as an assistant professor in 1966. He initially did research on surface science aspects (e.g., thermodynamics of surfaces, detergents, micelle formation, adsorption, Langmuir monolayers, and biophysics). During the early exploration and discovery stages of oil and gas in the North Sea, he became involved in the Danish Research Science Foundation programs, with other research institutes around Copenhagen, in oil recovery phenomena and surface science. Later, research grants on the same subject were awarded to him by European Union projects. These projects also involved extensive visits to other universities and an exchange of guests from all over the world. Dr. Birdi was appointed as a research professor in 1985 (Nordic Science Foundation), and was then appointed, in 1990, to the Danish Pharmacy University, Copenhagen, as a professor of physical chemistry. Since 1999, he has been actively engaged in consultancy for both industrial and university projects. There has been continuous involvement with various industrial contract research programs throughout these years. These projects have actually been a very important source of information in keeping up with real-world problems, and helped in the guidance of research planning at all levels. Dr. Birdi is a consultant to various national and international industries. He is and has been a member of various chemical societies, and a member of organizing committees of national and international meetings related to surface science. He has been a member of selection committees for assistant professors and professors, and was an advisory member (1985–1987) of the American Chemical Society journal *Langmuir*. Dr. Birdi has been an advisor for some 90 advanced student projects and various PhD projects. He has authored nearly 100 papers and articles (and a few 100 citations). In order to describe these research observations and data, he realized that it was essential to write books on the subject of surface and colloid chemistry. His first book on surface science was published in 1984 (*Adsorption and the Gibbs Surface Excess*, Chattorraj, D. K. and Birdi, K. S., Plenum Press, New York). This book still remains the only one of its kind in recent decades. Further publications include *Lipid and Biopolymer Monolayers at Liquid Interfaces* (K. S. Birdi, Plenum Press, New York, 1989); *Fractals, in Chemistry, Geochemistry and Biophysics* (K. S. Birdi, Plenum Press, New York, 1994); *Handbook of Surface and Colloid Chemistry* (first edition, 1997; second edition, 2003; third edition, 2009; CD-ROM 1999, CRC Press, Boca

Raton (Ed. K. S. Birdi); *Self-Assembly Monolayer* (Plenum Press, New York, 1999); and *Scanning Probe Microscopes* (CRC, Boca Raton, 2002a). Surface and colloid chemistry has remained his major interest of research.

1 Introduction to Surfaces and Colloids

1.1 INTRODUCTION

In everyday life, reactions and changes are observed that are dependent on the structures of the involved matter (which may be **solids, liquids,** or **gases**). However, in many industrial (chemical industry and technology) and natural biological phenomena, it is found that some processes require a more detailed definition of matter. Matter exists as

Gas
Liquid
Solid

phases, as has been recognized by classical science (as depicted in the following text).

1.2 SOLID PHASE–LIQUID PHASE–GAS PHASE

The molecules that are situated at the *interfaces* (e.g., between gas–liquid, gas–solid, liquid–solid, liquid$_1$–liquid$_2$, and solid$_1$–solid$_2$) are known to behave differently from those in the bulk phase (Adam, 1930; Aveyard and Hayden, 1973; Bakker, 1926; Bancroft, 1932; Partington, 1951; Davies and Rideal, 1963; Defay et al., 1966; Gaines, 1966; Harkins, 1952; Holmberg, 2004; Matijevic, 1969; Fendler and Fendler, 1975; Adamson and Gast, 1997; Chattoraj and Birdi, 1984; Birdi, 1989, 1997, 1999, 2002, 2009; Miller and Neogi, 2008; Somasundaran, 2006). Typical examples are

- Liquid surfaces
- Surfaces of oceans (liquid–air interface)
- Solid surfaces (adhesion, glues, tapes)
- Road surface (solid–air or solid–car tire)
- Lung surface
- Washing and cleaning (surfaces, foams)
- Emulsions (cosmetics, pharmaceutical products)
- Diverse industries (oil and gas, paper, milk products)

In this chapter, a very short general description of surfaces and colloids (small particles) is given. For instance, reactions taking place at the surface of the oceans will be expected to be different from those observed deep inside the water. Further, in some

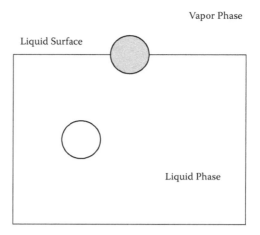

FIGURE 1.1 Surface molecules (schematic).

instances such as oil spills, the importance of the role of the ocean surface can be easily realized. It is also well known that the molecules situated near or at the interface (i.e., liquid–gas) will be interacting differently with respect to each other than the molecules in the bulk phase (Figure 1.1). This important aspect will be described extensively in this book. The intramolecular forces acting would thus be different in these two cases. In other words, all processes occurring near any interface will be dependent on these molecular orientations and interactions. Furthermore, it has been pointed out that, for a dense fluid, repulsive forces dominate the fluid structure and are of primary importance. The main effect of these repulsive forces is to provide a uniform background potential in which the molecules move as hard spheres. The molecules at the interface would be under an asymmetrical force field, which produces the so-called surface tension or interfacial tension (Figure 1.2) (Chattoraj and Birdi, 1984; Birdi, 1989, 1997, 1999, 2002; Adamson and Gast, 1997). This leads to the adhesion forces between liquids and solids, which is a major application area of surface and colloid science.

The resultant force on molecules will vary with time because of the movement of the molecules; the molecules at the surface will be pointed downward into the bulk

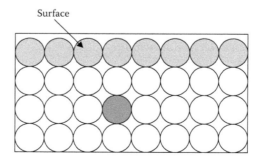

FIGURE 1.2 Intermolecular forces around a molecule in the bulk liquid (dark) and around a molecule in the surface (light) layer (schematic).

phase. The nearer the molecule is to the surface, the greater the magnitude of the force due to *asymmetry*. The region of asymmetry plays a very important role. Thus, when the surface area of a liquid is increased, some molecules must move from the interior of the continuous phase to the interface. The surface tension of a liquid is the force acting normal to the surface per unit length of the interface, thus tending to reduce the surface area. The molecules in the liquid phase are surrounded by neighboring molecules, and these interact with each other in a symmetrical way. In the gas phase, where the density is 1000 times lesser than in the liquid phase, the interactions between molecules are very weak as compared to those in the dense liquid phase. Thus, when we pass from the liquid phase to the gas phase, there is a change in density of factor 1000. This means that, while in the liquid phase, a molecule occupies a volume that is 1000 times smaller than when it is in the gas phase.

Surface tension is the differential change of free energy with change in surface area. An increase in surface area requires that molecules from the bulk phase are brought to the surface phase. This is valid when there are two fluids or a solid and a liquid, and it is usually designated *interfacial tension*.

A molecule of a liquid attracts the molecules that surround it, and, in its turn, it is attracted by them (Figure 1.2). For the molecules that are inside a liquid, the resultant of all these forces is neutral, and all of them are in equilibrium by reacting with each other. When these molecules are on the surface, they are attracted by the molecules below and by the lateral ones, but not toward the outside. The resultant is a force directed inside the liquid. In its turn, the cohesion among the molecules supplies a force tangential to the surface. So, a fluid surface behaves like an elastic membrane that wraps and compresses the liquid below. The surface tension expresses the force with which the surface molecules attract each other. It is common observation that, due to the surface tension, it takes some effort for some bugs to climb out of the water in lakes. On the contrary, other insects, such as the marsh treaders and water striders, exploit the surface tension to skate on the water without sinking (Figure 1.3).

Another well-known example is the floating of a metal needle (heavier than water) on the surface of water (Figure 1.4.) The surface of a liquid can thus be regarded as the plane of potential energy. It may be assumed that the surface of a liquid behaves as a membrane (at a molecular scale) that stretches across and needs to be broken in order to be penetrated. One observes this *tension* when considering that a heavy iron needle (heavier than water) can be made to float on the water surface when carefully placed (Figure 1.4).

The reason a heavy object floats on water is because in order to sink, it must overcome the surface forces. This clearly shows that, at any liquid surface, there exists a tension (*surface tension*) that needs to be broken when any contact is made between the liquid surface and the material (here the steel needle). A liquid can form three types of interfaces:

1. Liquid and vapor or gas (e.g., ocean surface and air)
2. Liquid$_1$ and liquid$_2$ immiscible (water–oil, *emulsion*)
3. Liquid and solid interface (water drop resting on a solid, wetting, cleaning of surfaces, adhesion).

FIGURE 1.3   Insect strides on the surface of water.

FIGURE 1.4   An iron needle floats on the surface of water (*only if carefully placed*; otherwise, it should sink due to gravity forces).

Solid surfaces can similarly exhibit additional characteristics, where solid and $solid_2$ interact (such as in: cement, adhesives, etc.)

1. $Solid_1$–$solid_2$ (cement, adhesives)

An analogous case would be when the solid is crushed and the surface area increases per unit gram (Figure 1.5). For example, finely divided talcum powder has a surface area of 10 m^2/g. Active charcoal exhibits surface areas corresponding to over 1000 m^2/g. This is obviously an appreciable quantity. Qualitatively, one must notice that work has to be put into the system when one increases the surface area (both for liquids or solids or any other interface).

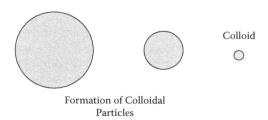

Formation of Colloidal
Particles

FIGURE 1.5 Formation of fine (colloidal) particles (schematic; size less than micrometer).

The surface chemistry of *small particles* is an important part of everyday life (such as dust, talcum powder, sand, raindrops, emissions, etc.). The designation *colloid* is used for particles that are of small dimension and that cannot pass through a membrane with a pore size ca. 10^{-6} m (= micrometer [µm]) (Thomas Graham described this about a century ago).

The nature and relevance of colloids is one of the main current research topics (Birdi, 2002). They are an important class of materials, intermediate between bulk and molecularly dispersed systems. Colloid particles may be spherical but, in some cases, one dimension can be much larger than the other two (as in a needle-like shape). The size of particles also determines whether they can be seen by the naked eye. Colloids are not visible to the naked eye or under an ordinary optical microscope. The scattering of light can be suitably used to see such colloidal particles (such as dust particles, etc.). Their size then may range from 10^{-4} to 10^{-7} cm. The units used are as follows:

$1 \text{ µm} = 10^{-6}$ m
$1 \text{ nm} = 10^{-9}$ m
$1 \text{ Å (Angstrom)} = 10^{-8} \text{ cm} = 0.1 \text{ nm} = 10^{-10}$ m

The unit Angstrom (Å) is related to the famous Swedish scientist, and currently nm (10^{-9} m) is mainly used. Nanosize (nanometer range) particles are currently of much interest in different applied-science systems. The word *nano* is derived from Greek and means *dwarf*. Nanotechnology has actually been getting a strong boost from the last decade of innovation, as reported by the surface and colloid literature. In fact, light scattering is generally used to study the size and size distribution of such systems. Since colloidal systems consist of two or more phases and components, the interfacial area-to-volume ratio becomes very significant. Colloidal particles have a high surface-area-to-volume ratio compared with bulk materials. A significant proportion of the colloidal molecules lie within, or close to, the interfacial region. Hence, the interfacial region has significant control over the properties of colloids. To understand why colloidal dispersions can either be stable or unstable, we need to consider the following:

1. The effect of the high surface-area-to-volume ratio (e.g., 1000 m² surface area per gram of solid [active charcoal])
2. The forces operating between the colloidal particles

There are some very special characteristics that must be considered as regards colloidal particle behavior: size and shape, surface area, and surface charge density. The Brownian motion of particles is a much-studied field. The fractal nature of surface roughness has recently been shown to be of importance (Birdi, 1993). Recent applications have been reported where *nanocolloids* have been employed. Therefore, some terms are needed to be defined at this stage. The definitions generally employed are as follows. *Surface* is a term used when one considers the dividing phase between

Gas–Liquid
Gas–Solid

Interface is the term used when one considers the dividing phase:

Solid–Liquid (Colloids)
$Liquid_1$–$Liquid_2$ (Oil–Water: emulsion)
$Solid_1$–$Solid_2$ (Adhesion: glue, cement)

In other words, surface tension may be considered to arise due to a degree of unsaturation of bonds that occurs when a molecule resides at the surface and not in the bulk. The term is used for solid–vapor or liquid–vapor interfaces. The term *interfacial tension* is more generally used for the interface between two liquids (oil–water), two solids, or a liquid and solid. It is, of course, obvious that, in a one-component system, the fluid is uniform from the bulk phase to the surface. However, the orientation of the surface molecules will be different from those in the bulk phase. For instance, in the case of water, the orientation of molecules inside the bulk phase will be different from those at the interface. Hydrogen bonding will orient the oxygen atom toward the interface. The question one may ask, then, is: how sharply does the density change from that of a fluid to that of gas (a change by a factor of 1000)? Is this transition region a monolayer deep or many layers deep? Many reports are found where this subject has been investigated.

The **Gibbs adsorption theory** (Birdi, 1989, 1999, 2002, 2008; Defay et al., 1966; Chattoraj and Birdi, 1984) considers the surface of liquids to be monolayer. The surface tension of water decreases appreciably on the addition of very small quantities of *soaps and detergents*. The Gibbs adsorption theory relates the change in surface tension to the change in soap concentration. The experiments that analyze the spread monolayers are also based on one molecular layer. The latter data indeed conclusively verifies the Gibbs assumption (as described later). Detergents (soaps, etc.) and other similar kind of molecules are found to exhibit self-assembly characteristics. The subject related to *self-assembly monolayer* (SAM) structures will be treated extensively (Birdi, 1999). However, no procedure exists that can provide information by direct measurement. The composition of the surface of a solution with two components or more would require additional comments.

Colloids (Greek for glue-like) are a wide variety of systems consisting of finely divided particles or macromolecules (such as glue, gelatin, proteins, etc.) that are found in everyday life. In Table 1.1 examples of such typical colloidal suspensions are given. Further, colloidal systems are widespread in their occurrence, and have

TABLE 1.1

Typical Colloidal Systems

Phases		
Dispersed	Continuous	System Name
Liquid	Gas	Aerosol fog, spray
Gas	Liquid	Foam, thin films, froth
		Fire extinguisher foam
Liquid	Liquid	Emulsion (milk) mayonnaise, butter
Solid	Liquid	Sols, AgI, photography films
		Suspension wastewater
		Cement, coal slurry
Biocolloids Corpuscles	Serum	Blood, blood-coagulants
Hydroxyapatite	Collagen	Bone
Liquid	Solid	Solid emulsion (toothpaste)
Solid	Gas	Solid aerosol (dust)
Gas	Solid	Solid foam—expanded (polystyrene)
		Insulating foam
Solid	Solid	Solid suspension/solids in plastics

biological and technological significance. There are three types of colloidal systems (Adamson and Gast, 1997; Birdi, 2002, 2008):

1. In simple colloids, a clear distinction can be made between the disperse phase and the disperse medium, for example, simple emulsions of oil-in-water (o/w) or water-in-oil (w/o).
2. Multiple colloids involve the coexistence of three phases, of which two are finely divided, such as multiple emulsions (mayonnaise, milk) of water-in-oil-in-water (w/o/w) or oil-in-water-in-oil (o/w/o).
3. Network colloids have two phases forming an interpenetrating network, for example, a polymer matrix.

Colloidal (as solids or liquid drops) stability is determined by the free energy (surface free energy or the interfacial free energy) of the system. The main parameter of interest is the large surface area exposed between the dispersed phase and the continuous phase. Since the colloid particles move about constantly, their dispersion energy is determined by *Brownian motion*. The energy imparted by collisions with the surrounding molecules at temperature T = 300 K is $3/2\ k_{BT} = 3/2\ 1.38\ 10^{-23}\ 300 = 10^{-20}$ J (where k_B is the Boltzmann constant). This energy and the intermolecular forces would thus determine colloidal stability.

In the case of colloid systems (particles or droplets), the kinetic energy transferred on collision will be thus $k_{BT} = 10^{-20}$ J. However, at a given moment, there is a high probability that a particle may have a larger or smaller energy. Further, the probability of total energy several times k_{BT} (over 10 times k_{BT}) thus becomes very small. Instability will be observed if the ratio of the barrier height to k_{BT} is around 1–2 units.

The idea that two species (solid–solid) should interact with one another, so that their mutual potential energy can be represented by some function of the distance between them, has been described in the literature. Furthermore, colloidal particles frequently adsorb (and even absorb) ions from their dispersing medium (such as in groundwater treatment and purification). Sorption that is much stronger than what would be expected from dispersion forces is called *chemisorption*, a process that is of both chemical and physical interest. For example, in a recent report, it was mentioned that finely divided iron particles could lead to enhanced photosynthesis in oceans (resulting in the binding of large amounts of CO_2). This could lead to control of the global warming effect.

Emulsions: As one knows from experience, the fact that oil and water do not mix suggests that these systems are dependent on the oil–water interface. The $liquid_1$–$liquid_2$ (oil–water) interface is found in many systems, and the most important is the world of *emulsions*.

The trick in using emulsions is that both *water* and *oil* (the latter is insoluble in water) can be applied simultaneously. Further, other molecules may be included that may be soluble in either phase (water or oil). This obviously leads to the observation of thousands of applications of emulsions. It is very important to mention here that, actually, nature uses this trick in most of the major biological fluids. The most striking example is milk. The emulsion chemistry of milk has been found to be the most complex, and still not very well investigated.

In fact, the state obtained by mixing oil and water is an important example of interfacial behavior of $liquid_1$–$liquid_2$. Emulsions of oil–water systems are useful in many aspects of daily life, such as milk, foods, paint, oil recovery, pharmaceutical, and cosmetics. The colloidal chemistry of milk makes it the most complicated naturally made product.

If olive oil is mixed with water and shaken, the following is obtained:

About 1 mm diameter oil drops are formed.
After a few minutes, the oil drops merge together and two layers are again formed.

However, if suitable substances are added that change surface forces, the olive oil drops formed can be very small (in the micrometer range).

These considerations are important in regard to different systems such as paints, cements, adhesives, photographic products, water purification, sewage disposal, emulsions, chromatography, oil recovery, paper and print industry, microelectronics, soaps and detergents, catalysts, and biological systems (such as cell, virus). In some oil–surfactant–water–diverse components, liquid crystal (LC) phases (lyotropic LC) are observed. These are indeed the basic building blocks in many applications of emulsions in technology.

2 Capillarity and Surface Forces (Liquids)

2.1 INTRODUCTION

It is common observation that a liquid takes the shape of a container that surrounds or contains it. However, it is also found that, in many cases, there are other subtle properties that arise at the interface of liquids. The most common behavior is bubble and foam formation. Another phenomena is that, when a glass capillary tube is dipped in water, the fluid rises to a given height. It is observed that the narrower the tube, the higher the water rises. The role of liquids and liquid surfaces is important in many everyday natural processes (e.g., oceans, lakes, rivers, raindrops, etc.). Therefore, in these systems, one will expect the surface forces to be important, considering that the oceans cover some 75% of the surface of the earth. Accordingly, there is a need to study surface tension and its effect on surface phenomena in these different systems. This means that the structures of molecules in the bulk phase need to be considered in comparison to those at the surface.

The *surface molecules* are under a different force field from the molecules in the bulk phase or the gas phase. These forces are called *surface forces*. A liquid surface behaves like a stretched elastic membrane in that it tends to contract. This action arises from the observation that, when one empties a beaker with a liquid, the liquid breaks up into spherical drops. This phenomenon indicates that drops are being created under some forces that must be present at the surface of the newly formed interface. These surface forces become even more important when a liquid is in contact with a solid (such as ground–water; oil reservoir). The flow of liquid (e.g., water or oil) through small pores underground is mainly governed by *capillary forces*. Capillary forces are found to play a very dominant role in many systems, which will be described later. Thus, the interaction between liquid and any solid will form curved surface that, being different from a planar fluid surface, initiates the capillary forces.

In this chapter, the basics of surface forces will be described, and examples will be given where the system is dependent on these forces. The principles of surface forces are the building blocks that lead to the understanding of the subject. These forces interact at both the $liquid_1$–$liquid_2$ and liquid–solid interfaces.

2.2   SURFACE FORCES (IN LIQUIDS)

It is important to consider the molecular interactions in liquids that are responsible for their physicochemical properties (such as boiling point, melting point, heat of vaporization, surface tension, etc.), which enables one to both describe and relate the different properties of matter in a more clear manner (both qualitatively and quantitatively). These ideas form the basis for *quantitative structure activity relationship* (QSAR; Birdi, 2002). This approach toward analysis and application is becoming more common due to the enormous help available from computers.

The different kinds of forces acting between any two molecules are dependent mainly on the distance between the two molecules. The difference in distance between molecules in liquid or gas can be estimated as follows. In the case of water, the following data are known:

Water data
    Volume per mole **liquid** water = V_{liquid} = 18 mL/mol
    Volume per mole water in **gas** state (at STP) (V_{gas}) = 22 L/mol

One thus finds (in general) that the ratio V_{gas}/V_{liquid} = ca. 1000. Hence, the approximate distance between molecules will be ca. $10 = (1000)^{1/3}$ (from simple geometrical considerations of volume [% length^3] and length). In other words, the density of water changes 1000 times as the surface is crossed from the liquid phase to the gas phase (Figure 2.1). This large change means that the surface molecules must be under an environment different from that the liquid or gas phase. The distance between gas molecules is approximately 10 times larger than in a liquid. Hence, the forces (*all forces increase when distances between molecules decrease*) between gas molecules is much weaker than in the case of the liquid phase. All interaction forces between molecules (solid phase, liquid phase, gas phase) are related to the distance between molecules.

It is the cohesive forces that maintain water, for example, in the liquid state at room temperature and pressure. This becomes obvious when one compares two different molecules, such as H_2O and H_2S. At room temperature and pressure, H_2O is

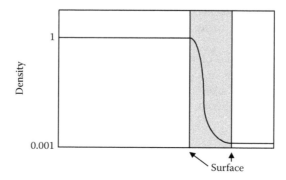

FIGURE 2.1   Change of density of a fluid (water) near the surface.

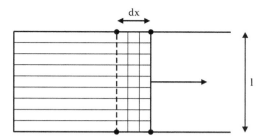

FIGURE 2.2 Surface film of a liquid (see text for details).

a liquid, while H_2S is a gas. This means that H_2O molecules interact with different forces to form a liquid phase. On the other hand, H_2S molecules exhibit much lower interactions and thus are in a gas phase at room temperature and pressure.

2.2.1 SURFACE ENERGY

The state of *surface energy* has also been described by the following classic example (Adamson and Gast, 1997; Chattoraj and Birdi, 1984; Birdi, 1989; Birdi, 1997, 2002). Consider the area of a liquid film that is stretched in a wire frame by an increment dA, whereby the surface energy changes by (γ dA) (Figure 2.2). Under this process, the opposing force is **f**. From these data on dimensions, we find that

Surface tension = γ
Change in area = $dA = l\,dx$
Change in x-direction = dx

$$\mathbf{f}\,dx = \gamma\,dA \tag{2.1}$$

or

$$\gamma = \mathbf{f}\,(dx/dA)$$

$$= \mathbf{f}/2\,l \tag{2.2}$$

where dx is the change in displacement, and l is the length of the thin film. The quantity γ represents the force per unit length of the surface (mN/m = dyn/cm), and this force is defined as surface tension or interfacial tension. The *surface tension*, γ, is the differential change of free energy with change of surface area at constant temperature, pressure, and composition.

One may consider another example to describe surface energy. Let us imagine that a liquid fills a container of the shape of a funnel. In the funnel, if one moves the liquid upwards, then there will be an increase in *surface area*. This requires that some molecules from the bulk phase have to move into the surface area and create extra surface A_s. The work required to do so will be (force × area) γA_s. This work is reversible at constant temperature and pressure, and thus gives the increase in free energy of the system:

$$dG = -\gamma A_S \tag{2.3}$$

Thus, the tension per unit length in a single surface, or surface tension γ, is numerically equal to the surface energy per unit area. Then G_s, the surface free energy per unit area is

$$G_s = \gamma = (d\,G/d\,A) \tag{2.4}$$

Under reversible conditions, the heat (**q**) associated with it gives the surface entropy S_S:

$$d\mathbf{q} = T\,dS_S \tag{2.5}$$

Combining these equations we find that

$$d\gamma/dT = -Ss \tag{2.6}$$

Further, we find that

$$H_S = G_S + T_S \tag{2.7}$$

and we can also write for surface energy E_S:

$$E_S = G_S + T_S\,S \tag{2.8}$$

These relations give

$$E_S = \gamma - T\,(d\gamma/dT) \tag{2.9}$$

The quantity E_S has been found to provide more useful information on surface phenomena than any of the other quantities.

Thus, S_s is the surface entropy per square centimeter of surface. This shows that, to change the surface area of a liquid (or solid, as described in later text), there exists a *surface energy* (γ: *surface tension*) that needs to be considered.

The quantity γ means that, to create 1 m² (= 10^{20} Å²) of new surface of water, one will need to use 72 mJ energy. To transfer a molecule of water from the bulk phase (where it is surrounded by about 10 near neighbors by about 7 k_B T; k_B T = 4.12 10^{-21}J) to the surface, about half of these hydrogen bonds need to be broken (i.e., 7/2 k_B T = 3.5 k_B T). The free energy of transfer of one molecule of water (with area of 12 Å²) will be thus about 10^{-20} J (or about 3 k_B T). This is a reasonable quantity under these assumptions.

Further similar consideration is needed if one increases the surface area of a solid (e.g., by *crushing*). In the latter case, one needs to measure and analyze the surface tension of the solid. It is found that energy needed to crush a solid is related to the surface forces (i.e., solid surface tension). More examples as given later will provide real-world situations where γ of both liquids and solids is needed in order to describe the physics of the surfaces.

2.3 LAPLACE EQUATION (LIQUID CURVATURE AND PRESSURE)

Let us consider aspects in the field of wettability. Surely everybody has noticed that water tends to rise near the walls of a glass container. This happens because the molecules of this liquid have a strong tendency to adhere to glass. Liquids that wet the walls make concave surfaces (e.g., water/glass), and those that do not wet them make convex surfaces (e.g., mercury/glass). Inside tubes with internal diameter smaller than 2 mm, called capillary tubes, a wettable liquid forms a concave meniscus in its upper surface and tends to go up along the tube. On the contrary, a nonwettable liquid forms a convex meniscus and its level tends to go down. The amount of liquid attracted by the capillary rises until the forces that attract it balance the weight of the fluid column. The rising or lowering of the level of liquids into thin tubes is named *capillarity* (*capillary force*). Further, capillarity is driven by the forces of cohesion and adhesion already mentioned.

One notices that a liquid inside a large beaker is almost *flat* at the surface. However, the same liquid inside a fine tubing will be found to be *curved* (Figure 2.3). This behavior is of much importance in everyday life. The physical nature of this phenomena will be the subject of this section.

The surface tension, γ, and the mechanical equilibrium at interfaces have been described in the literature in detail (Adamson and Gast, 1997; Chattoraj and Birdi, 1984; Birdi, 1989, 2002, 2008). The surface has been considered as a hypothetical stretched *membrane*, which is termed as the surface tension. In a real system undergoing an infinitesimal process, it can be written that

$$dW = p \, dV + p' \, dV' - \gamma \, dA \qquad (2.10)$$

where dW is the work done by the systems when a change in volume dV and dV', occurs; p and p' are pressures in the two phases α and β, respectively, at equilibrium; and dA is the change in the interfacial area. The sign of the interfacial work is designated as negative by convention (Chattoraj and Birdi, 1984).

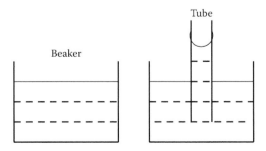

FIGURE 2.3   Surface of water inside a large beaker and in a narrow tubing.

The fundamental property of liquid surfaces is that they tend to contract to the smallest possible area. This property is observed in the spherical form of small drops of liquid, in the tension exerted by soap films as they tend to become less extended, and in many other properties of liquid surfaces. In the absence of gravity effects, these curved surfaces are described by the Laplace equation, which relates the mechanical forces as (Adamson and Gast, 1997; Chattoraj and Birdi, 1984; Birdi, 1997):

$$p - p' = \gamma \left(1/r_1 + 1/r_2 \right) \tag{2.11}$$

$$= 2 \left(\gamma/r \right) \tag{2.12}$$

where r_1 and r_2 are the radii of curvature (in the case of an ellipse), while r is the radius of curvature of a spherical-shaped interface. It is a geometric fact that surfaces for which Equation 2.11 hold are surfaces of minimum area. These equations thus give

$$dW = p \, d(V + V') - \gamma \, d\mathbf{A} \tag{2.13}$$

$$= p \, dV^t - \gamma \, d\mathbf{A} \tag{2.14}$$

where $p = p'$ for a plane surface, and V^t is the total volume of the system.

It will be shown here that, due to the presence of surface tension in liquids, a pressure difference exists across the curved interfaces of liquids (such as drops or bubbles). This *capillary force* will be analyzed later.

If one dips a tubing into water (or any fluid) and applies a suitable pressure, then a bubble is formed (Figure 2.4). This means that the pressure inside the bubble is greater than the atmospheric pressure. Thus, curved liquid surfaces induce effects that need special physicochemical analyses in comparison to flat liquid surfaces. It must be noted that, in this system, a mechanical force has induced a change on the

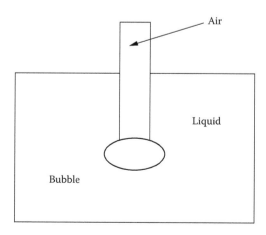

FIGURE 2.4   Formation of an air bubble in a liquid.

surface of a liquid. The phenomena is also called *capillary force*. One may ask then if a similar consideration is also required in the case of solids. It has been found that the answer to the latter is positive. For example, in order to remove a liquid that is inside a porous media such as a sponge, force equivalent to these capillary forces would be needed.

As seen in Figure 2.4, the bubble is produced by applying a suitable pressure, ΔP, to obtain a bubble of radius R, where the surface tension of the liquid is γ. Let us consider a situation the bubble is expanded by applying pressure P_{inside}. The surface area of the bubble will increase by dA, and the volume by dV. This means that there are two opposing actions: expansion of volume and of area. The work done can be expressed in terms of that done against the forces of surface tension and that done in increasing the volume. At equilibrium, the following condition will exist between these two kinds of work:

$$\gamma \, dA = (P_g - P_{liquid}) \, dV \tag{2.15}$$

where $dA = 8 \pi R \, dR$ ($A = 4 \pi R^2$), and $dV = 4 \pi R^2 \, dR$ ($V = 4/3 \pi R^3$). Combining these relations gives the following:

$$\gamma \, 8 \pi R \, dR = \Delta P \, 4 \pi R^2 \, dR \tag{2.16}$$

and

$$\Delta P = 2 \gamma / R \tag{2.17}$$

where $\Delta P = (P_g - P_{liquid})$. Since the free energy of the system at equilibrium is constant, $dG = 0$, and then these two changes in the system are equal. If the same consideration is applied to the soap bubble, then the expression for ΔP bubble will be

$$\Delta P_{bubble} = P_{inside} - P_{outside}$$
$$= 4 \gamma / R \tag{2.18}$$

because now there exists *two* surfaces, and the factor **2** is needed to consider this state.

The pressure applied produces work on the system, and the creation of the bubble leads to the creation of a surface area increase in the fluid. The Laplace equation relates the pressure difference across any curved fluid surface to the curvature, 1/**radius** and its surface tension γ. In those cases where nonspherical curvatures are present, the more universal equation is obtained:

$$\Delta P = \gamma (1/R_1 + 1/R_2) \tag{2.19}$$

It is also seen that, in the case of spherical bubbles, since $R_1 = R_2$, this equation becomes identical to Equation 2.18. Thus, in the case of a liquid drop in air (or gas

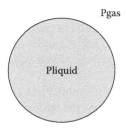

FIGURE 2.5 Liquid drop and $\Delta \mathbf{P}$ (= $P_{LIQUID} - P_{GAS}$) (at a curved liquid–gas interface).

phase), the Laplace pressure would be the difference between the pressure inside the drop, p_L, and the gas pressure, p_G (Figure 2.5):

$$p_L - p_G = \Delta \mathbf{P} \tag{2.20}$$

$$= 2 \ \gamma/\text{radius} \tag{2.21}$$

In the case of a water drop with radius 2 μ, there will be $\Delta \mathbf{P}$ of magnitude

$$\Delta \mathbf{P} = 2 \ (72 \ \text{mN/m})/ \ 2 \ 10^{-6} \ \text{m}$$

$$= 72000 \ \text{Nm}^{-2} = 0.72 \ \text{atm} \tag{2.22}$$

It is known that the difference between pressure inside the drop and the gas pressure changes the vapor pressure of a liquid. Thus, $\Delta \mathbf{P}$ will affect the vapor pressure and lead to many consequences in different systems.

The Laplace equation is useful for analyses in a variety of systems:

1. Bubbles or drops (raindrops or combustion engines, sprays, fog)
2. Blood cells (flow of blood cells through arteries)
3. Oil or groundwater movement in rocks
4. Lung vesicles

2.3.1 THE IMPORTANT ROLE OF LAPLACE EQUATION IN VARIOUS APPLICATIONS

Another important conclusion one may draw is that $\Delta \mathbf{P}$ is larger inside a *small* bubble than in a *larger* bubble with the same γ. This means that, when bubbles meet, the smaller bubble will enter the larger bubble to create a new bubble (Figure 2.6.) This phenomenon will have much important consequences in various systems (such as emulsion stability, lung alveoli, oil recovery, bubble characteristics [as in champagne; beer]). The same consequences will be observed when two liquid drops contact each other: the smaller will merge into the larger drop.

Figure 2.7 shows a system that initially shows two bubbles of different curvature. After the tap is opened, the smaller bubble is found to shrink, while the larger bubble (with lower $\Delta \mathbf{P}$) increases in size until equilibrium is reached (when the curvature of the two bubbles become equal in magnitude). This type of equilibrium is the

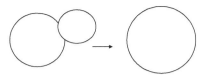

FIGURE 2.6   Coalescence of two bubbles with different radii.

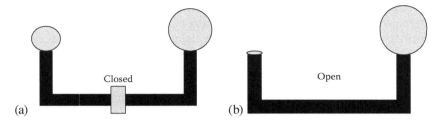

FIGURE 2.7a,b   Equilibrium state of two bubbles of different radii (see text).

basis of lung alveoli, in which fluids (containing lipid surfactants) balance out the expanding–contracting cycle.

It has been observed that a system with varying size bubbles collapses faster than when bubbles (or liquid drops) are of exactly the same size. Another major consequence is observed in the oil recovery phenomena. Oil production takes place (in general) by applying gas or water injection. When gas or water injection is applied where there are small pores, the pressure needed will be higher than that in the large-pore zone. Thus, the gas or water will *bypass* the small-pore zone and leave the oil behind (at present more than 30% to 50% of the oil in place is not recovered under normal production methods). This obviously presents a great challenge to surface and colloid chemists in the future. Enhanced oil recovery (EOR) investigations will be described later in this book (Chapter 6.2).

The capillary (Laplace) pressure determines many industrial and biological systems. The lung alveoli are dependent on the radii during the inhale–exhale process, and the change in surface tension of the fluid lining them. In fact, many lung diseases are related to the lack of surface pressure and capillary pressure balance. Blood flow through arteries of different diameter throughout the body is another system where the Laplace pressure is of much interest for analytical methods (such as heart function and control).

In industry, the magnitude of surface tension can be monitored with the help of bubble pressure. Air bubbles are pumped through a capillary into the solution, and the pressure measured is calibrated to known surface tension solutions. Using a suitable computer, one can then estimate surface tension values very accurately. Commercial apparatus are also available to monitor surface tension.

The consequence of Laplace pressure is very important in many different processes. One example is that, when a small drop comes into contact with a large drop, the former will merge into the latter. Another aspect is that vapor pressure over a curved liquid surface, p_{cur}, will be larger than on a flat surface, p_{flat}. A relation between pressure over curved and flat liquid surfaces was derived (Kelvin equation):

$$\ln (p_{cur}/p_{flat}) = (v_L/\mathbf{R}\ \mathbf{T}) (2\ \gamma/R_{cur}) \qquad (2.23)$$

where p_{cur} and p_{flat} are the vapor pressures over curved and flat surfaces, respectively. R_{cur} is the radius of curvature, and v_L is the molar volume. The Kelvin equation thus suggests that, if liquid is present in a porous material, such as cement, then the difference in vapor pressure exits between two pores of different radii due to the high vapor pressure. Similar consequences of vapor press exists when two solid crystals of different size are considered. The smaller-sized crystal will exhibit higher vapor pressure and will also result in a faster solubility rate.

Further, the transition from water vapor in clouds to rain drops is not as straightforward as it might seem. The formation of a large liquid raindrop requires that a certain number of water molecules in the clouds form a nuclei. The nuclei or embryo will grow, and the Kelvin relation will be the determining factor.

2.4   CAPILLARY RISE (OR FALL) OF LIQUIDS

The behavior of liquids in narrow tubes is one of the most common examples in which capillary forces are involved. It will be shown later how important this phenomenon is in many different parts of everyday life and technology. In fact, liquid curvature is one of the most important physical surface properties that requires attention in most of the application areas of this science. The range of these applications is from blood flow in the veins to oil recovery in the reservoir. Properties of fabrics are also governed by capillary forces (i.e., wetting, etc.). The sponge absorbs water or other fluids where the capillary forces push the fluid into the many pores of the sponge. This is also called *wicking* process (as in candlewicks).

Let us analyze the system in which a narrow capillary circular tube is dipped into a liquid. The liquid is found to rise in the capillary when it wets the capillary (like water and glass or water and metal). The *curvature* of the liquid inside the capillary will lead to pressure difference between this state and the relatively flat surface outside the capillary (Figure 2.8). The fluid with surface tension, γ, wets the inside of the tubing, which brings about an equilibrium of capillary forces. However, if the fluid is nonwetting (such as Hg in glass), then one finds that the fluid *falls*. This arises because Hg does not wet the tube. Capillary forces arise from the difference in attraction of the liquid molecules to each other and the attraction of the liquid molecule to those of the capillary tube. The fluid rises inside the narrow tube to a height **h** until the surface tension forces balance the weight of the fluid. This equilibrium gives following relation:

$$\gamma\, 2\ \pi\ \mathbf{R} = \text{surface tension force} \qquad (2.24)$$

$$= \rho_L\ g_g\ \mathbf{h}\ \pi\ \mathbf{R}^2 = \text{fluid weight} \qquad (2.25)$$

where γ is the surface tension of the liquid, and R is the radius of curvature. In the case of narrow capillary tubes (less than 0.5 mm), the curvature can be safely set equal to the radius of the capillary tubing. The fluid will rise inside the tube to compensate for surface tension force, and thus, at equilibrium, we get

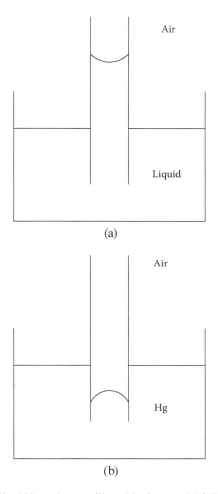

(a)

(b)

FIGURE 2.8 Rise of liquid in a glass capillary (a) of water, (b) fall of Hg.

$$\gamma = 2 \, \mathbf{R} \, \rho_L \, \mathbf{g} \, \mathbf{h} \qquad\qquad (2.26)$$

where ρ_L is the density of the fluid, $\mathbf{g}$ is the acceleration of gravity, and $\mathbf{h}$ is the rise in the tube. Water ($\gamma = 72.8$ mN/m, $\rho_L = 1000$ kg/m^3; $\mathbf{g} = 9.8$ m/s^2) will rise to a height of

0.015 mm in a capillary of radius 1 m
1.5 mm in a capillary of radius 1 cm
14 cm in a capillary of radius 0.1 mm

If the radius of the capillary is larger than 0.5 mm, then one will need corrections for accurate measurements. Tables for such corrections are published in the current literature (Adamson and Gast, 1997). This phenomenon is of importance for plants with very high stems in which water is needed for growth. However, in some particular case where the contact angle, θ, is not zero, a correction will be needed, and the equation will become

$$\gamma = 2\ \mathbf{R}\ \rho_L\ g\ \mathbf{h}\ (1/\cos{(\theta)}) \tag{2.27}$$

It is seen that, when the liquid wets the capillary wall, the magnitude of is 0, and $\cos(0) = 1$. In the case of Hg, the contact angle is $180°$ since it is a nonwetting fluid (see Figure 2.8). Because $\cos(180) = -1$, the sign of $\mathbf{h}$ in the equation will be negative. This means that Hg will show a drop in height in a glass tubing. Hence, the rise or fall of a liquid in a tubing will be governed by the sign of $\cos{(\theta)}$. Thus, capillary forces will play an important factor in all systems where liquids are present in a porous environment.

Similar results can also be derived by using the Laplace equation (Equation 2.21) (1/radius = 1/R):

$$\Delta P = 2\ \gamma/\mathbf{R} \tag{2.28}$$

The liquid rises to a height $\mathbf{h}$ and the system achieves equilibrium, and the following relation is found:

$$2\ \gamma/\mathbf{R} = \mathbf{h}\ g_g\ \rho_L \tag{2.29}$$

This can be rewritten as

$$\gamma = 2\ \mathbf{R}\ \rho_L\ g_g\ \mathbf{h} \tag{2.30}$$

The various surface forces are seen to be responsible for capillary rise. The lower the surface tension, the lower the height of the column in the capillary. The magnitude of γ is determined from the measured value $\mathbf{h}$ for a fluid with known ρ_L. The magnitude of $\mathbf{h}$ can be measured directly by using a suitable device (e.g., a photograph image).

Further, it is known that real-world capillaries or pores are not always circular shaped. In fact, in oil reservoirs, the pores are more triangular shaped or square shaped than circular. In this case, the rise in capillaries of other shapes, such as rectangular or triangular (Birdi et al., 1988; Birdi, 1997, 2002) can be measured. These studies have much significance in oil recovery or water treatment systems. In any system in which the fluid flows through porous material, it would be expected that capillary forces would be one of the most dominant factors.

Also, the vegetable world is known to be dependent on capillary pressure (and osmotic pressure) to bring water up to the higher parts of plants. Using these forces, some trees succeed in bringing the essential liquid (water) up to 120 m above the ground.

2.5   BUBBLE FORMATION

There is one phenomenon that everyone can quickly recognize—soap bubbles—that one has observed since childhood. The formation of foam bubbles along the coasts of lakes and oceans is also another common experience. Scientists have investigated bubble formation extensively (Boys, 1959; Lovett, 1994; Birdi, 1997). It is also known that soap bubbles are extremely thin and unstable. In spite of the latter, under special conditions, soap bubble can be kept intact for long periods of time, which thus allows one to study its physical properties (such as thickness, composition, conductivity, spectral reflection, etc.).

The thickness of a bubble is, in most cases, over hundreds of micrometer in the initial state. The film consists of a bilayer of detergent that contains the solution. The film thickness decreases with time due to following reasons:

Drainage of fluid away from the film
Evaporation

Therefore, the stability and lifetime of such thin films will be dependent on these different characteristics. This is evident from the fact that, as an air bubble is blown under the surface of a soap or detergent solution, it will rise up to the surface. It may remain at the surface if the speed is slow, or it may escape into the air as a soap bubble. Experiments show that a soap bubble consists of a very thin liquid film with an iridescent surface. But, as the fluid drains away and the thickness decreases, the bubble approaches the equivalent of barely two surfactant molecules plus a few molecules of water. It is worth noting that the limiting thickness is of the order of *two or more surfactant molecules*. This means that one can see with the naked eye the molecular-size structures of thin liquid films (TLFs) (if curved).

As the air bubble enters the surface region, the soap molecules are pushed up and, as the bubble is detached, it leaves as a TLF with the following characteristics (as found from various measurements):

A bilayer of soap (approximately 200 Å thick) on the outer region contains the aqueous phase.
The thickness of the initial soap bubble is of a certain micrometer.
The thickness decreases with time, and one starts to observe rainbow colors as the reflected light is of the same wavelength as the thickness of the bubble (a few hundred Angstroms).
The thinnest liquid film consist mainly of the bilayer of surface-active substance (such as soap = 50 Å) and some layers of water. The light interference and reflection studies show many aspects of these TLFs.

The iridescent colors of the soap bubble arise from the interference of reflected light waves. Reflected light from the outer surface and the inner layer cause this interference effect (Figure 2.9). Rainbow colors are observed as the bubble thickness decreases due to the evaporation of water. Thicker films reflect red light, and blue–green colors are observed. Lesser thin films cancel out the yellow wavelength, and blue color is observed. As the thickness approaches the wavelength of light, all colors are canceled out and a *black (or gray) film* is observed. This corresponds to 25 nm (250 Å).

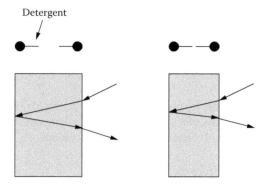

FIGURE 2.9   Reflection of light waves from the outer layer and the inner layer. Interference is caused by the difference in the path of light traveling inside the thin film.

The transmitted light, I_{tr}, is related to the incident light, I_{in}, and the reflected intensity, I_{re}:

$$I_{tr} = I_{in} - I_{re} \qquad (2.31)$$

The daylight consists of different wavelengths of colors:

Red	680 nm
Orange	590 nm
Yellow	580 nm
Green	530 nm
Blue	470 nm
Violet	405 nm

Slightly thicker soap films (ca. 150 nm) sometimes look golden. In the thinning process, the different colors get cut off. Thus, if blue color gets cut off, the film appears amber to magenta.

The following are some recipes for soap bubble solutions:

Soap bubble solutions: In general one may use a detergent solution (1–10 g/L detergent concentration). However, to produce stable bubbles (which means slow evaporation rates, stability to vibration effects, etc.), some additives have been used. A common recipe is

Detergent (dishwashing)	1–10 g/L
Glycerin	Less than 10%
Water	The rest

Another recipe has the following composition (Lovett, 1994):

100 g glycerine + 1.4 g triethanolamine + 2 g oleic acid

Bubble film stability can be described as follows:

Evaporation of water
Flow of water away from the film
Stability of the bubble film

Evaporation can be reduced by containing the bubble in a closed bottle. It is also found that, in such a closed system, the bubbles remain stable for a very long time. The drainage of water away from the film is dependent on the viscosity of the fluid. Therefore, such additives as glycerin (or other thickening agents [polymers]) assist in maintaining stability.

2.6 MEASUREMENT OF SURFACE TENSION OF LIQUIDS

It is thus apparent from what has been discussed so far that the measurement of the surface tension of liquids is an important analysis. The method to use in the measurement of γ depends on the system and experimental conditions (as well as the accuracy needed). For example, if the liquid is water (at room temperature), then the method used will be different from that if the system is molten metal (at very high temperature, ca. 500°C). These different systems will be explained in the methods described in this section.

2.6.1 LIQUID DROP WEIGHT AND SHAPE METHOD

The formation of liquid drops when flow occurs through thin tubes is a common daily phenomenon. In some cases, such as eyedrops, the size of the drop plays a significant role in the application and dosage of the medicine. The drop formed when liquid flows through a circular tube is shown in Figure 2.10.

In many processes (such as oil recovery, blood flow, underground water), one encounters liquid flow through thin (micrometer diameter), noncircular-shaped tubes, or pores. In the literature, one finds studies that address these latter systems. In another context of liquid drop formation, for example, in an inkjet nozzle, this technique falls under a class of scientifically challenging technology. The inkjet printer demands such quality that this branch of drop-on-demand technology is much in the realm of industrial research. All combustion engines are controlled by oil drop formation and evaporation characteristics. The important role of capillary forces is obvious in such systems.

As the liquid drop grows larger, it will, at some stage, break off the tube (due to gravity force being larger than the surface force) and will correspond to the maximum weight of the drop that can hang. The equilibrium state where the weight of the drop is exactly equal to the detachment surface energy is given as

$$m_m \, \mathbf{g} = 2 \, \pi \, \mathbf{R} \, \gamma \qquad (2.32)$$

where m_m is the weight of the detached drop, and $\mathbf{R}$ is the radius of the tubing.

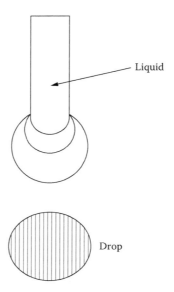

FIGURE 2.10   Drop formation of a liquid in a tube

A simple method is to *count* the number of drops (e.g., 10 or more) and measure the weight.

A more convenient method may be used, in which a fluid is pumped and the drops are collected and weighed. Since in some systems (solutions) there may be kinetic effects, care must be taken to keep the flow as slow as possible. This system is very useful in studying what one finds in daily life phenomena: oil flow, blood cells flowing through arteries, etc. In those cases where the volume of fluid available is limited, this method may be used with advantage. By decreasing the diameter of the tubing, one can work with fluids less than 1 mL. This may be the case for systems such as eye fluids, etc.

The magnitude of γ can be determined from either the maximum weight or the shape of the drop.

Maximum weight method: The "detachment" method is based upon the following: to detach a body from the surface of a liquid that wets the body, it is necessary to over-come the same surface tension forces that operate when a drop is broken away. The liquid attached to the solid surface on detachment creates the following surfaces:

     Initial stage: Liquid attached to solid
     Final stage: Liquid separated from solid

In the process from the *initial* to the *final* stage, the liquid molecules that were near the solid surface have been moved away and are now near their own molecules. This requires energy, and the force required to make this happen is proportional to the surface area of contact and the surface tension of the liquid.

Methods of determining surface tension by measuring the force required to detach a body from a liquid are therefore similar to the stalagmometer method described earlier.

Profile of a Pendant Drop

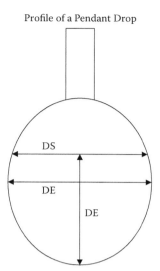

FIGURE 2.11 Pendant drop of liquid (shape analysis).

However, the advantage of the former over the latter method consists in that it makes it possible to choose the most convenient form and size of the body (platinum rod, ring, or plate) so as to enable the measurement to be carried out rapidly but without any detriment to its accuracy. The "detachment" method has found an application in the case of liquids whose surface tensions change with time.

Shape of the liquid drop (Pendant drop method): The liquid drop forms as it flows through a tubing (Figure 2.11). At a stage just before it breaks off, the shape of the *pendant drop* has been used to estimate γ. The drop shape is photographed and, from the diameter of the shape, γ can be accurately determined.

The parameters needed are as follows. A quantity pertaining to the ratio of two significant diameters are

$$\mathbf{S} = d_s/d_e \tag{2.33}$$

where d_e is the equatorial diameter, and d_s is the diameter at a distance d_e from the tip of the drop (Figure 2.11). The relation between γ and d_e and S is found as

$$\gamma = \rho_L \, \mathbf{g} \, d_e^2/\mathbf{H} \tag{2.34}$$

where ρ_L is the density of the liquid, and **H** is related to **S**, but the values of $1/\mathbf{H}$ for varying **S** were obtained from experimental data. For example, when $\mathbf{S} = 0.3$, $1/\mathbf{H} = 7.09837$, whereas, when $\mathbf{S} = 0.6$, then $1/\mathbf{H} = 1.20399$. Accurate mathematical functions have been used to estimate $1/\mathbf{H}$ for a given d_e value (Adamson and Gast, 1997; Birdi, 2002). The accuracy (0.1%) is satisfactory for most of the systems, especially when experiments are carried out under extreme conditions (such as high temperatures and pressures).

The pendant drop method is very useful under specific conditions:

1. Technically, only a drop (a few microliters) is required. For example, eye fluid can be studied since only a drop of one microliter is needed.
2. It can be used under very extreme conditions (very high temperature or corrosive fluids).
3. It can be used under very high pressure and temperatures. Oil reservoirs are found typically at 100°C and 300 atm pressure. The surface tension of such systems can be conveniently studied by using *high pressure and temperature* cells with optical clear windows (sapphire windows 1 cm thick; up to 2000 atm). For example, γ of inorganic salts at high temperatures (ca. 1000°C) can be measured using this method. The variation in surface tension can be studied as a function of various parameters (temperature and pressure; additives [gas, etc.]).

2.6.2 THE RING METHOD

A method that has been rather widely used involves the determination of the force to detach a ring or loop of wire from the surface of a liquid. Originally developed by du Nouy, this method is based on using a ring (platinum) and measuring the force when it is dipped in the liquid surface.

This is one of the many detachment methods of which the drop weight and the Wilhelmy slide methods are also examples. As with all detachment methods, one supposes that, within an accuracy of a few percent, the detachment force is given by the surface tension multiplied by the periphery of the surface (liquid surface) detached (from a solid surface of a tubing or ring or plate). This assumption is also found to be acceptable for most experimental purposes. Thus, for a ring, as illustrated in Figure 2.12,

$$W_{total} = W_{ring} + \mathbf{2} \, (2 \, \pi \, \mathbf{R}_{ring}) \, \gamma \tag{2.35}$$

$$= W_{ring} + 4 \, \pi \, R_{ring} \, \gamma \tag{2.36}$$

where W_{total} is the total weight of the ring, W_{ring} is the weight of the ring in air, and R_{ring} is the radius of the ring. The circumference is $2 \, \pi \, R_{ring}$, and factor $\mathbf{2}$ is because of the two sides of contact.

This relation assumes that the contact between the fluid and the ring is geometrically simple. It is also found that this relation is fairly correct (better than 1%) for most working situations. However, it was observed that Equation 2.36 needed a correction factor in much the same way as was done for the drop weight method. Here, however, there is one additional variable, so that the correction factor $\mathbf{f}$ now depends on two dimensionless ratios.

Experimentally, the method is capable of good precision. A so-called chainomatic balance has been used to determine the maximum pull, but a popular simplified version of the tensiometer, as it is sometimes called, makes use of a torsion wire and is quite compact. Among experimental details to mention are that the dry weight of the

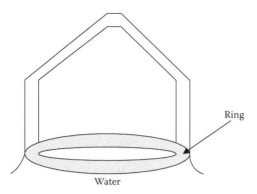

FIGURE 2.12 Ring method (see text for details).

ring, which is usually constructed of platinum, is to be used; the ring should be kept horizontal (a departure of 1° was found to introduce an error of 0.5%, whereas one of 2.1° introduced an error of 1.6%); and care must be taken to avoid any disturbance of the surface as the critical point of detachment is approached. The ring is usually flamed before use to remove surface contaminants such as grease, and it is desirable to use a container for the liquid that can be overflown so as to ensure the presence of a clean liquid surface.

A zero or near-zero contact angle is necessary; otherwise, results will be low. This was found to be the case with surfactant solutions in which adsorption on the ring changed its wetting characteristics, and in which liquid–liquid interfacial tensions were measured. In such cases, a Teflon or polyethylene ring may be used. When used to study monolayers, it may be necessary to know the increase in area at detachment, and some calculations of this are available (Adamson and Gast, 1997).

2.6.3 WILHELMY SLIDE (OR PLATE) METHOD

The methods so far discussed have required more or less tabular solutions, or otherwise correction factors to the respective "*ideal*" equations. Further, if continuous measurements need to be made, then it is not easy to use some of these methods (such as the capillary rise or bubble method).

The most useful method of measuring surface tension is by the well-known Wilhelmy plate method. If a plate-shaped metal is dipped in a liquid, the surface tension forces will be found to produce a tangential force (Figure 2.13). This is because a new contact phase is created between the plate and the liquid.

The total weight measured, W_{total}, would be

$$W_{total} = \text{weight of the plate} + \gamma \,(\text{perimeter}) - \text{updrift} \qquad (2.37)$$

The surface force will act along the perimeter of the plate (i.e., length $[L_p]$ + width $[W_p]$). The plate is often very thin (less than 0.1 mm) and made of platinum, but even plates made of glass, quartz, mica, and filter paper can be used. The forces acting on the plate consist of the gravity and surface tension downward, and buoyancy due

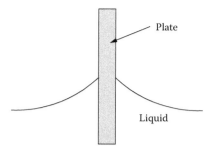

FIGURE 2.13 Wilhelmy plate in a liquid (plate with dimensions: length = Lp, width = Wp).

to displaced water upward. For a rectangular plate of dimensions L_p and W_p and of material density ρp, immersed to a depth h_p in a liquid of density ρL, the net downward force, F, is given by the following equation (i.e., weight of plate + surface force [$\gamma \times$ perimeter of the plate − upward drift]):

$$F = \rho p \; \mathbf{g} \; (Lp \; Wp \; tp) + 2 \; \gamma \; (tp + Wp)(\cos(\theta)) - \rho L \; \mathbf{g} \; (tp \; Wp \; hp) \qquad (2.38)$$

where γ is the liquid surface tension, θ is the contact angle of the liquid on the solid plate, and $\mathbf{g}$ is the gravitational constant. The weight of the plate is constant and can be tared. If the plate used is very thin (i.e., tp $\ll$ Wp) and the updrift is negligible (i.e., hp is almost zero), then γ is found from

$$\gamma = (F)/(2 \; Wp) \qquad (2.39)$$

The sensitivity of γ using these procedures has been found to be very high (± 0.001 dyn/cm [mN/m]) (Birdi, 2008). The change in surface tension (surface pressure = Π) is then determined by measuring the change in F for a stationary plate between a clean surface and the same surface with a monolayer present. If the plate is completely wetted by the liquid (i.e., $\cos(\theta) = \cos(0) = 1$), the surface pressure is then obtained from the following equation:

$$\Pi = -[\Delta F/2(tp + Wp)] = -\Delta F/2 \; Wp \; , \; \text{if Wp} \gg \text{tp} \qquad (2.40)$$

Thus, by using very thin plates with thickness 0.1 to 0.002 mm, one can measure surface tension with very high sensitivity. In practice, by using very thin platinum plates of well-known dimension (length = 1.00 or 2.00 cm), one can calibrate the apparatus with pure liquids, such as water and ethanol. The buoyancy correction is made very small (and negligible) by using a very thin plate and dipping the plate as little as possible.

The wetting of water on a platinum plate is achieved very satisfactorily by using commercially available plates that are roughened. This property produces almost complete wetting; that is, $\theta = 0$.

Sensitivity can thus be increased by using a very thin plate. The force is, in this way, determined by measuring the changes in the mass of the plate, which is directly coupled to a sensitive electrobalance.

2.7 SURFACE TENSION OF LIQUIDS

Surface tension of liquids has been extensively analyzed in the literature. Some typical values of surface tension of different liquids are given in Table 2.1. A brief analysis of these data is given in the following text.

Some comments are needed on these data in order to explain the differences in surface tension data and molecular structure. The range of γ is found to vary from ca. 20 to over 1000 Nm/m. The surface tension of Hg is high because it is a liquid metal with a very high boiling point. This indicates that it needs much energy to break the bonds between Hg atoms to evaporate. Similarly, γ of NaCl as a liquid (at high temperature) is also very high. The same case is found for metals in liquid state.

The other liquids can be considered as under each type, which should help understand the relation between the structure of a molecule and its surface tension.

Alkanes: The magnitude increases by 1.52 mN/m per two CH_2 when the alkyl chain length increases from 10 to 12 (*n*-Decane = 23.83; *n*-Dodecane = 25.35).

n-Heptane	20.14
n-Hexadecane	27.47
n-Hexane	18.43
n-Octane	21.62

Alcohols: The magnitude of γ changes by $23.7 - 22.1 = 1.6$ mN/m per $-CH_2-$ group. This is based upon the γ data of ethanol (22.1 mN/m) and propanol (23.7 mN/m). These observations indicate the molecular correlation between bulk forces and surface forces (surface tension γ) for homologous series of substances.

2.7.1 EFFECT OF TEMPERATURE AND PRESSURE ON SURFACE TENSION OF LIQUIDS

All natural processes are found to be dependent on the temperature and pressure effects on any system under consideration. For example, oil reservoirs are generally found under high temperature (ca. 100°C) and pressure (over 200 atm). Actually, humans are aware of the great variations in both temperature (sun) and pressure (earthquakes) with which natural phenomena surround the earth. Even the surface of the earth itself comprises temperature variation of −50°C to +50°C. On the other hand, the center mantle of the earth increases in temperature and pressure as one goes from the surface to the center of the earth (about 5000 km). Surface tension is related to the internal forces in the liquid (surface), and one must thus expect it to bear a relationship to internal energy. Further, it is found that surface tension always decreases with increasing temperature.

Surface tension, γ, is a quantity that can be measured accurately and applied in the analyses of all kinds of surface phenomena. If a new surface is created, then, in the case of a liquid, molecules from the bulk phase must move to the surface. The work required to create the extra surface area, d**A**, will be given as

$$dG = \gamma \, d\mathbf{A} \qquad (2.41)$$

TABLE 2.1
Surface Tension Values of Some Common Liquids

Liquid	Surface Tension
1,2-Dichloro-ethane	33.3
1,2,3-Tribromo propane	45.4
1,3,5-Trimethylbenzene (Mesitylene)	28.8
1,4-Dioxane	33.0
1,5-Pentanediol	43.3
1-Chlorobutane	23.1
1-Decanol	28.5
1-Nitro propane	29.4
1-Octanol	27.6
Acetone (2-Propanone)	25.2
Aniline	43.4
2-Aminoethanol	48.9
Anthranilic acid ethylester	39.3
Anthranilic acid methylester	43.7
Benzene	28.9
Benzylalcohol	39.0
Benzylbenzoate (BNBZ)	45.9
Bromobenzene	36.5
Bromoform	41.5
Butyronitrile	28.1
Carbon disulfide	32.3
Quinoline	43.1
Chlorobenzene	33.6
Chloroform	27.5
Cyclohexane	24.9
Cyclohexanol	34.4
Cyclopentanol	32.7
p-Cymene	28.1
Decalin	31.5
Dichloromethane	26.5
Diiodomethane (DI)	50.8
1,3-Diiodopropane	46.5
Diethylene glycol	44.8
Dipropylene glycol	33.9
Dipropylene glycol monomethylether	28.4
Dodecyl benzene	30.7
Ethanol	22.1
Ethylbenzene	29.2
Ethylbromide	24.2
Ethylene glycol	47.7
Formamide	58.2

TABLE 2.1 (*Continued*)
Surface Tension Values of Some Common Liquids

Liquid	Surface Tension
Fumaric acid diethylester	31.4
Furfural (2-Furaldehyde)	41.9
Glycerol	64.0
Ethylene glycol monoethylether (Ethyl cellosolve)	28.6
Hexachlorobutadiene	36.0
Iodobenzene	39.7
Isoamylchloride	23.5
Isobutylchloride	21.9
Isopropanol	23.0
Isopropylbenzene	28.2
Isovaleronitrile	26.0
m-Nitrotoluene	41.4
Mercury	425.4
Methanol	22.7
Methyl ethyl ketone (MEK)	24.6
Methylnaphthalene	38.6
N,N-dimethyl acetamide (DMA)	36.7
N,N-dimethyl formamide (DMF)	37.1
N-methyl-2-pyrrolidone	40.7
n-Decane	23.8
n-Dodecane	25.3
n-Heptane	20.1
n-Hexadecane	27.4
n-Hexane	18.4
n-Octane	21.6
n-Tetradecane	26.5
n-Undecane	24.6
n-Butylbenzene	29.2
n-Propylbenzene	28.9
Nitroethane	31.9
Nitrobenzene	43.9
Nitromethane	36.8
o-Nitrotoluene	41.5
Perfluoroheptane	12.8
Perfluorohexane	11.9
Perfluorooctane	14.0
Phenylisothiocyanate	41.5
Phthalic acid diethylester	37.0
Polyethylen glycol 200 (PEG)	43.5
Polydimethyl siloxane	19.0

Continued

TABLE 2.1 (*Continued*)
Surface Tension Values of Some Common Liquids

Liquid	Surface Tension
Propanol	23.7
Pyridine	38.0
3-Pyridylcarbinol	47.6
Pyrrol	36.6
sym-Tetrabromoethane	49.7
tert-Butylchloride	19.6
sym-Tetrachloromethane	26.9
Tetrahydrofuran (THF)	26.4
Thiodiglycol	54.0
Toluene	28.4
Tricresylphosphate (TCP)	40.9
Water	72.8
o-Xylene	30.1
m-Xylene	28.9
a-Bromonaphthalene (BN)	44.4
a-Chloronaphthalene	41.8
Mercury	425.4
Metals (liquid state: high temperature)	Greater than 1000

Note: Surface tension (20°C; mN/m [dyn/cm]).

The surface free energy, G_S, per unit area is given as

$$G_S = \gamma = (dG/dA)_{T,P} \tag{2.42}$$

Hence, the other thermodynamic surface quantities will be
Surface entropy:

$$S_S = -(dG_S/dT)_P \tag{2.43}$$

$$= -(d \, \gamma/dT) \tag{2.44}$$

We can thus derive for surface enthalpy H_S

$$H_S = G_S + T \, S_S \tag{2.45}$$

All natural processes are dependent on the effect of temperature and pressure. For instance, oil reservoirs are found under high temperatures (ca. 80°C) and pressure (around 100 to 400 atm [depending on the depth]). Thus, one must investigate such systems under these parameters. This is related to the fundamental equation for free energy, G, enthalpy, H, and entropy, S, of the system:

$$G = H - T S \qquad (2.46)$$

It is thus important that, in all practical analyses, one should be aware of the effects of temperature and pressure. The molecular forces that stabilize liquids will be expected to decrease as temperature increases. Experiments also show that, in all cases, surface tension decreases with increasing temperature.

The surface entropy of liquids is given by $(-d\ \gamma/dT)$. This means that the entropy is positive at higher temperatures. The rate of decrease of surface tension with temperature is found to be different for different liquids (Appendix A), which supports the foregoing description of liquids. This observation explains the molecular description of surface tension.

For example, the surface tension of water data is given as

At 5°C, $\gamma = 75$ mN/m
At 25°C, $\gamma = 72$ mN/m
At 90°C, $\gamma = 60$ mN/m

Extensive γ data for water was fitted to the following equation (Birdi, 2002):

$$\gamma = 75.69 - 0.1413\ t_C - 0.0002985\ t_C^2 \qquad (2.47)$$

where t_C is in degree Centigrade. This equation gives the value of γ at 0°C as 75.69 mN/m. The value of γ at 50°C is found to be $(75.69 - 0.1413 \times 50 - 0.0002985 \times 50 \times 50) = 60$ mN/m.

Further, these data show that γ of water decreases with temperature from 25°C to 60°C, $(72 - 60)/(90 - 25) = 0.19$ mN/mC. The differences in surface entropy yields information on the structures of different liquids. It is also observed that the effect of temperature will be lower for liquids with higher boiling point (such as Hg) than for low-boiling liquids (such as n-hexane). Actually, a correlation exists between and heat of vaporization (or boiling point). In fact, many systems even show big differences when comparing winter or summer months (such as raindrops, sea waves, foaming in natural environments, etc.).

Different thermodynamic relations have been derived that can be used to estimate surface tension at different temperatures. In this regard, straight-chain alkanes have been especially extensively analyzed. The data show that there exists a simple correlation among surface tension, temperature, and n_C, which allows one to estimate the value of surface tension of a given alkane at any temperature. This observation has many aspects in applied industry. It allows one to quickly estimate the magnitude of surface tension of an alkane at the required temperature. Further, one has fairly good quantitative analysis of how surface tension will change for a given alkane under a given experimental condition. A detailed analysis is given in Appendix A.

2.7.2 Different Aspects of Liquid Surfaces

2.7.2.1 Heat of Surface Formation and Evaporation

In order to understand how liquids are stabilized, several attempts have been made to relate the surface tension of a liquid to the latent heat of evaporation. It was argued a century ago (Stefan, 1886) that, when a molecule is brought to the surface of a liquid from the interior, the work done in overcoming the attractive force near the surface should be related to the work expended when it escapes into the scarce vapor phase (Adamson and Gast, 1997; Birdi, 1997, 2002). It was suggested that the first quantity should be approximately half of the second. According to the Laplace theory of capillarity force, the attractive force acts only over a small distance equal to the radius of the sphere (see Chapter 1, Figure 1.1) and, in the interior, the molecule is attracted equally in all directions and experiences no resultant force. On the surface, it experiences a force due to the liquid in the hemisphere, and half the total molecular attraction is overcome in bringing it there from the interior.

Accordingly (Stefan, 1886), the energy needed to bring a molecule from the bulk phase to the surface of a liquid should be *half* the energy needed to bring it entirely into the gas phase. From geometrical considerations, one knows that a sphere can be surrounded by 12 molecules of the same size, which corresponds with the most densely packed structure. The distance between gas molecules is approximately 10 times greater than that in liquids or solids. This indicates that intermolecular forces in liquids would be weaker than that in solids by a few orders of magnitude, as is also found experimentally.

The ratio of the enthalpy of surface formation to the enthalpy of vaporization, $h_s{:}h_{vap}$, for various substances is given in Table 2.2. Substances with nearly spherical-shaped molecules have ratios near 1/2, while substances with a polar group on one end have a much smaller ratio. This difference indicates that the latter molecules are oriented with the nonpolar end toward the gas phase and the polar end toward

TABLE 2.2
Enthalpy of Surface Formation, hs
[10 erg/molecule], and Ratio to
Enthalpies of Vaporization, hvap

Molecules (Liquid)	hs/hvap
Hg	0.64
N_2	0.51
O_2	0.5
CCl_4	0.45
C_6H_6	0.44
Diethyl ether	0.42
$Cl\text{-}C_6H_5$	0.42
Methyl formate	0.40
Ethyl acetate	0.4
Acetic acid	0.34
H_2O	0.28
C_2H_5OH	0.19
CH_3OH	0.16

Note: See text for details.

the liquid phase. In other words, molecules with dipoles would be expected to be oriented perpendicularly at the gas/liquid interfaces.

In fact, any deviation from Stefan's law is an indication that the surface molecules are oriented differently from those in the bulk phase. This observation is useful in order to understand surface phenomena.

As an example, one may proceed with this theory and estimate the surface tension of a liquid with data on its heat of evaporation. The number of near neighbors of a surface molecule will be about half (6 = 12/2) than those in the bulk phase (12 neighbors). It is now possible to estimate the ratio of the attractive energies in the bulk and surface phases per molecule. We have the following data for a liquid such as CCl_4:

$$\text{Molar energy of vaporization} = \Delta U_{vap} \tag{2.48}$$

$$= \Delta h_{vap} - RT$$
$$= 34000 \text{ Jmol}^{-1} - 8.315 \text{ J K}^{-1} \text{ mol}^{-1} (298 \text{ K})$$
$$= 31522 \text{ J mol}^{-1} \tag{2.49}$$

$$\text{Energy change per molecule} = 31522 \text{ Jmol}^{-1}/6.023 \ 10^{23} \text{ mol}^{-1} = 5.23 \ 10^{-20} \text{ J} \tag{2.50}$$

If we assume that about half of the energy is gained when a molecule is transferred to the surface, we get

$$\text{Energy per molecule at surface} = 5.23 \ 10^{-20} \ (2) \text{ J} = 2.6 \ 10^{-20} \text{ J} \tag{2.51}$$

The molecules at the surface occupy a certain value of area, which can be estimated only roughly as follows:

Density of CCl_4 = 1.59 g cm^{-3}
Molar mass = 12 + 4 (35.5) = 154 g mol^{-1}
Volume per molecule = 154/1.59 = 97 cm^3 mol^{-1}
Volume per molecule = 97 10^{-6} m^3 mol^{-1}/6.023 10^{23} mol^{-1}
= 1.6 10^{-28} m^3

The radius of a sphere (volume = 4/3 Π R^3) with this magnitude of volume =
    [1.6 10^{-28}/(4/3 Π)]$^{1/3}$
= 3.5 10^{-10} m

Area per molecule = Π R^2
= Π (3.5 10^{-10})^2 = 38 10^{-20} m^2

Surface tension (calculated) for CCl_4 = 2.6 10^{-20} J/38 10^{-20} m^2
= 0.068 J m^{-2} = 68 mNm^{-1}

The measured value of γ of CCl_4 is 27 mN/m (Table 2.1). The large difference can be ascribed to the assumption that Stefan's ratio of 2 was used in this example.

2.7.2.2 Other Surface Properties of Liquids

Surface waves on liquids: A liquid surface, for example, that of oceans or lakes, exhibit wave formation when strong winds are blowing over it. It is known that such waves are created by wind energy being transposed to waves. Hence, humans have tried to convert wave energy to other useful forms of energy. Both transverse capillary waves and longitudinal waves can deliver information about the elasticity and viscosity of surfaces, albeit on very different time scales. Rates of adsorption and desorption can also be deduced. Transverse capillary waves are usually generated with frequencies between 100 and 300 Hz. The generator is a hydrophobic knife edge situated on the surface and oscillating vertically, while the usual detector is a lightweight hydrophobic wire lying on the surface parallel to the generator edge. The generator and detector are usually close together (15–20 mm) so that reflections set up a pattern of standing transverse wave. The damping of capillary ripples arises primarily from the compression and expansion of the surface and the interaction between the surface film and propagation at the top of the path, and in the reverse direction at the bottom (Adamson and Gast, 1997).

This leads to compression and expansion of the surface. If a surface film is present, compression tends to lower the surface tension, whereas expansion raises it. Then a Marangoni flow is generated, which opposes the wave motion and dampens it (Birdi, 2002, 2009). Furthermore, if the material is soluble, the compression–expansion cycle will be accompanied by capillary ripples according to the hydrodynamic theories that have been described earlier. For longitudinal waves, a barrier (usually on the trough of a surface film) generate the length of the trough.

2.7.2.3 Interfacial Tension of Liquid₁–Liquid₂

It is a well-known adage that oil and water do not mix. However, it will be shown that, by changing the interfacial forces at the oil–water boundary, one can indeed *disperse* oil in water (or vice versa). At the oil–water interface there exists interfacial tension (IFT), which can be measured by some of the methods mentioned earlier (e.g., by drop weight, pendant drop, or Wilhelmy plate).

The interfacial tension γ_{AB} between two liquids with surface tension γ_A and γ_B is of interest in such systems as emulsions and wetting (Adamson and Gast, 1997; Chattoraj and Birdi, 1984; Somasundaran, 2006). An empirical relation was suggested (Antonow's rule), by which one can predict the surface tension γ_{AB}:

$$\gamma_{AB} = |\gamma_{A(B)} - \gamma_{B(A)}| \qquad (2.52)$$

The prediction of γ_{AB} from this rule is approximate but found to be useful in a large number of systems (such as alkanes:water), with some exceptions (such as water:butanol) (Table 2.3). For example,

TABLE 2.3
Antonow's Rule and Interfacial Tension Data [mN/m]

Oil Phase	w(o)	o(w)	o/w	w(o)–o
Benzene	62	28	34	34
Chloroform	52	27	23	24
Ether	27	17	8	9
Toluene	64	28	36	36
n-Propylbenze	68	29	39	40
n-Butylbenzne	69	29	41	40
Nitrobenzene	68	43	25	25
i-Pentanol	28	25	5	3
n-Heptanol	29	27	8	2
CS$_2$	72	52	41	20
Methyleneiodide	72	51	46	22

Note: See text for details.

$$\gamma_{water} = 72 \text{ mN/m (at 25°C)},$$

$$\gamma_{hexadecane} = 20 \text{ mN/m (at 25°C)}$$

$$\gamma_{water-hexadecane} = 72 - 20 = 52 \text{ mN/m (measured} = 50 \text{ mN/m)} \qquad (2.53)$$

However, for general considerations, one may only use it as a reliable guideline and when exact data are not available.

Antonow's rule can be understood in terms of a simple physical picture. There should be an adsorbed film or Gibbs monolayer of substance B (the one of lower surface tension) on the surface of liquid A. If we regard this film as having the properties of bulk liquid B, then $\gamma_{A(B)}$ is effectively the interfacial tension of a duplex surface and would be equal to $[\gamma_{A(B)} + \gamma_{B(A)}]$.

2.7.2.4 Measurement of Interfacial Tension (between Two Immiscible Liquids)

IFT can be measured by different methods, depending on the characteristics of the system. The following are some of them:

Wilhelmy plate method
Drop weight method (can also be used for high pressure and temperature)
Drop shape method (can also be used for high pressure and temperature)

The Wilhelmy plate is placed at the surface of the water, and the oil phase is added until the whole plate is covered by the latter. The apparatus must be calibrated with known IFT data such as water–hexadecane (52 mN/m; 25°C) (Table 2.4).

TABLE 2.4
Interfacial Tensions IFT between
Water and Organic Liquids (20°C)

Water/Organic Liquid	IFT [mN/m]
n-Hexane	1.0
n-Octane	50.8
CS_2	48.0
CCl_4	45.1
$Br-C_6H_5$	38.1
C_6H_6	35.0
$NO_2-C_6H_5$	26.0
Ethyl ether	10.7
n-Decanol	10
n-Octanol	8.5
n-Hexanol	6.8
Aniline	5.9
n-Pentanol	4.4
Ethyl acetate	2.9
Isobutanol	2.1
n-Butanol	1.6

The drop weight method is carried out by using a pump to deliver the water phase into the oil phase (or vice versa, as one finds suitable). The water drops sink to the bottom of the oil phase. The weight of the drops is measured (by using an electrobalance), and IFT can be calculated. Accuracy can be very high by choosing the right kind of setup. The drop shape (pendant drop) is most convenient if small amounts of fluids are available and extreme temperatures and pressures are involved. Modern digital image analyses also makes this method very easy to apply in extreme situations.

2.8 APPLICATIONS

The role of surface tension of liquids is found in many important systems. The capillary vise is found to play an important role in many everyday processes (such as washing and cleaning, plants, etc.).

3 Surfactants (Soaps and Detergents) and Physicochemical Properties

3.1 INTRODUCTION

As known from experience, any physical property of a liquid will change when a substance (called *solute*) is dissolved in it. Of course, the change may be small or large, depending on the concentration of the solute and other parameters. Accordingly, the magnitude of surface tension of a liquid will change when a solute is dissolved in it. This is to be expected from physicochemical considerations. It also becomes apparent that, if one could manipulate the surface tension of water, then many application areas would be affected drastically. In this chapter, some important *surface-active substances* having such properties will be described. These are specific substances that are used to change the surface tension of water in order to apply this characteristic to some useful purpose in everyday life.

The magnitude of surface tension change will depend on the concentration and the solute added. In some cases, the surface tension (γ) of the solution (such as NaCl) increases. The change in γ may be small (per mole added) (as in the case of inorganic salts) or large (as in the case of such molecules as ethanol or other soap-like molecules) with the addition of solute (equal gram per liter):

Inorganic salt: *Minor change* (increase) in γ
Ethanol or similar: *Small change* (decrease) in γ
Soap or similar: *Large change* (decrease) in γ

The following are typical surface tension data of different solutions:

Surface tension(mN/m)	72	50	40	30	22
Surfactant ($C_{12}H_{25}SO_4Na$)	0	0.0008	0.003	0.008	—
Ethanol	0	10	20	40	100

This shows that, to reduce the value of γ of water from 72 to 30 mN/m, one would need to add 0.005 moles of sodium dodecyl sulfate (SDS) or 40% ethanol. Of course, these two solutions cannot be used for the same application based on their similar magnitudes of γ.

There are special substances called *soaps* or *detergents* or *surfactants* that exhibit unique physicochemical properties. The most significant structure of these molecules is manifested by the presence of a *hydrophobic* (alkyl group) and a *hydrophilic* (polar groups, such as -OH, -CH$_2$CH$_2$O-, -COONa, -SO$_3$Na, -SO$_4$Na, -CH$_{33}$N-, etc.). The different polar groups are

Ionic groups
   Negatively charged—*anionic*
      -COONa
      -SO$_3$Na
      -SO$_4$Na
      -(N)(CH$_3$)$_4$Br (positively charged—*cationic*)
      -(N)(CH$_3$)$_2$-CH$_2$-COONa (Amphoteric)
Nonionic groups
   -CH$_2$CH$_2$OCH$_2$CH$_2$OCH$_2$CH$_2$OH
   -(CH$_2$CH$_2$OCH$_2$CH$_2$O) × (CH$_2$CH$_2$CH$_2$O)yOH

Accordingly, one also calls these substances *amphiphiles* (meaning two kinds, i.e., the alkyl part and the polar group):

CCCCCCCCCCCCCCCCCCCC-**O**
ALKYL GROUP(*CCCCCCC*-)—POLAR GROUP(-**O**) = AMPHIPHILE …
   (CH$_3$CH$_2$CH$_2$CH$_2$CH$_2$CH$_2$CH$_2$CH$_2$CH$_2$)—POLAR

In representations, the alkyl group is depicted as (.............), while the polar group is depicted as 0.

For instance, surfactants dissolve in water and give rise to low surface tension even at very low concentrations (a few grams per liter or 1–100 mmol/L) of the solution; therefore, these substances are also called surface-active molecules (*surface-active agents* or substances). On the other hand, most inorganic salts increase the surface tension of water. All surfactant molecules are amphiphilic, which means that these molecules exhibit hydrophilic and hydrophobic properties. Ethanol reduces the surface tension of water, but one will need over a few moles per liter to obtain the same reduction as when using a few millimoles of surface-active agents.

In comparison, addition of organic molecules such as methanol or ethanol decreases the magnitude of γ of water from 72 mN/m rather slowly. The value of γ decreases from 72 to 22 mN/m in pure ethanol. In comparison, the value of γ of surfactant solutions decreases to 30 mN/m with surfactant concentration around millimoles per liter (range of 1 to 10 g/L). Soaps have been used by humans for many centuries. In biology, one finds a whole range of amphiphile molecules (bile salts, fatty acids, cholesterol and other related molecules, phospholipids).

Surfactants are among the most important substances that play an essential role in everyday life. There also exist many surfactants in nature (such as bile acids in the stomach) that behave exactly the same way as human-made surface-active agents.

The γ of proteins (which are large molecules with molecular weights from 6000 to one million) also decrease when dissolved in water.

Surfactants are characterized as amphiphiles. Amphiphile in Greek means "likes both kinds." A part of the amphiphile likes oil or the hydrophobic (lipophilic) part, while the other part likes water or the hydrophilic (also called *lipophobic*) part. The balance between hydrophilic and lipophilic is called HLB. It can be estimated by experimental means, and theoretical analyses allow one to estimate its value (Adamson and Gast, 1997; Holmberg, 2004; Birdi, 2009). HLB values are applied in the emulsion (soaps and detergents) industry. Soap molecules are made by reacting fats with strong alkaline solutions (a process called *saponification*). In water solution, the soap molecule, $C_nH_{2n+1}COONa$ (with n greater than 12 to 22), dissociates at high pH into $RCOO^-$ and Na^+ ions.

Originally, humans used soaps produced from fats (especially lard). However, many decades ago, synthetic surfactants were made (as by-products of oil refining) for special industrial applications such as cleaning and washing. A great variety of surfactants were synthesized from oil by-products, especially, $C_{12}H_{25}C_6H_4SO_3Na$, and sodium dodecyl benzene sulfonates were used in detergents. Later, they were replaced by sodium dodecyl sulfates or sulfonates because sodium dodecyl benzene is not biodegradable. This means that the bacteria in sewage plants were not able to degrade the alkyl group. Alkyl sulfonates were degraded to alkyl hydroxyls and alkyl aldehydes, and later to CO_2, etc. This problem was solved by using biodegradable alkyl sulfates. In many applications, it has been found necessary to employ surfactants that were nonionic, a variety of which have been synthesized, and one can obtain tailor-made surfactants that suit a particular application. Further, since nonionic detergents do not exhibit any charges, these find applications in processes in which this property is essential.

For example, nonionic detergents as used in washing clothes are much different in structure and properties compared to those used in dishwashing machines. In washing machines, foam is crucial as it helps keep the dirt, once it has been removed, away from the clothes. On the other hand, dishwashing machines do not need any foam, as it will hinder the dishwashing process. However, in these cases, surface tension (which means wetting and other properties at the interface) needs to be low to decrease the contact angle as well as to remove fats (through detergent action). There are tailor-made nonionic detergents that satisfy these criteria. Another important property of a surface-active agent is that most of the ionic detergents cannot be used in conjunction with seawater (due high content of Ca and Mg salts). Therefore, special detergents have been used to combat this problem.

3.2   SURFACE TENSION OF AQUEOUS SOLUTIONS

The surface tension (γ) of any pure liquid (water or organic liquid) will change when another substance (solute) is dissolved in it. The change in γ will depend on the characteristics of the solute added. The surface tension of water increases (in general) when inorganic salts (such as NaCl, KCl, Na_2SO_4, etc.) are added (Figure 3.1), while its value decreases when organic substances are dissolved (ethanol, methanol, fatty acids, soaps, detergents, etc.) (Figure 3.1).

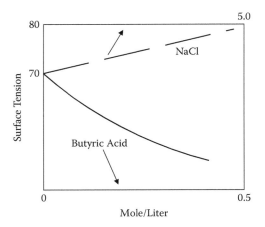

FIGURE 3.1   Change in surface tension of water as a function of added solutes (inorganic salts such as NaCl, surface-active agents such as butyric acid).

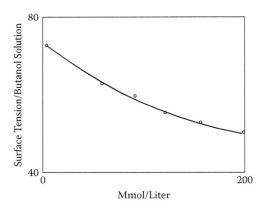

FIGURE 3.2   Surface tension plot of n-butanol solutions.

The surface tension of water increases from 72 to 73 mN/m when 1 M NaCl is added. On the other hand, the magnitude of surface tension decreases from 72 to 39 mN/m when only 0.008 M (0.008 M × 288 = 2.3 g/L) SDS (mol. wt. = 288) is dissolved. It thus becomes obvious, that in all those systems in which surface tension plays an important role, the additives will become significant. The data for n-butanol solutions in water are shown to decrease from 72 mN/m (pure water) to 50 mN/M in 200 mmol/L (Figure 3.2).

The γ of water changes slowly as compared to detergent solutions. Methanol–water mixtures gave the following γ data (at 20°C):

%w Methanol	0	10	25	50	80	90	100
γ		72	59	46	35	27	25	22.7

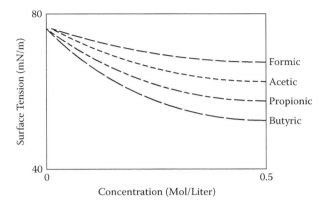

FIGURE 3.3   Surface tension data of a homologous series of short-chain acids in water.

The surface tension data in the case of a homologous series of alcohols and acids show a simple relation to the alkyl chain length (Figure 3.3). It is observed that each addition of the -CH_2- group gives values of concentration and surface tension such that the value of concentration is lower by about a *factor 3*.

However, it should be noted that such dependence in the case of nonlinear alkyl chains will be different. The effective -CH_2- increase in the case of nonlinear chains will be lesser (ca. 50%) than in the case of a linear alkyl chain. The tertiary -CH_2-group effect would be even lesser. In general, though, one will expect that the change in γ per mole substance will increase with any increase in the hydrocarbon group of the amphiphile.

The effect of chain length on surface tension arises from the fact that, as the hydrophobicity increases with each -CH_2- group, the amphiphile molecule adsorbs more at the surface. This will thus be a general trend in more complicated molecules also, such as proteins and other polymers. In proteins, the amphiphilic property arises from the different kinds of amino acids (25 different amino acids). Some amino acids have lipophilic groups (such as phenylalanine, valine, leucine, etc.), while others have hydrophilic groups (such as glycine, aspartic acid, etc.) (Figure 3.4).

In fact, one finds from surface tension measurements that some proteins are considerably more hydrophobic (such as hemoglobin) than others (such as bovine serum albumin, ovalbumin). These properties of proteins have been extensively investigated.

3.2.1   Surface-Active Substances (*Amphiphiles*)

All molecules that, when dissolved in water, reduce surface tension are called *surface-active substances* (e.g., soaps, surfactants, detergents). This means that such substances adsorb at the surface and reduce surface tension. The same will happen if a surface-active substance is added to a system of oil–water. The interfacial tension of the oil–water interface will be reduced accordingly. Inorganic salts, on the other hand, increase the surface tension of water.

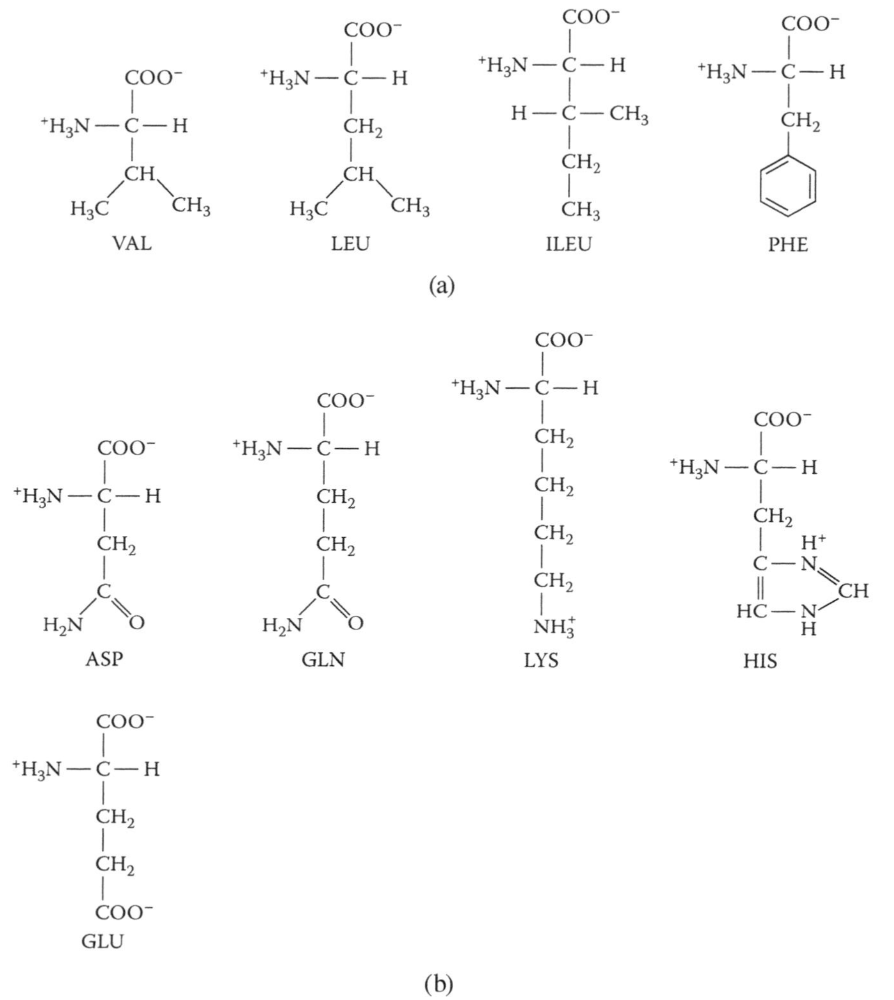

(a)

(b)

FIGURE 3.4　Polar (a) and apolar (b) amino acids.

Surfactants exhibit surface activity, which means these molecules will adsorb preferentially at interfaces:

Air–water
Oil–water
Solid–water

The magnitude of surface tension is reduced since the hydrophobic alkyl chain or group is energetically more attracted to the surface than being surrounded by water molecules inside the bulk aqueous phase. Figure 3.5 shows the monolayer formation of the surface-active substance at a high bulk concentration. Since the closely packed

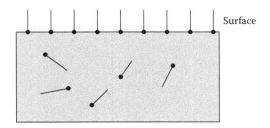

FIGURE 3.5 Orientation of soap (surface-active substance) at the surface of water (alkyl group:----, polar group: 0).

surface-active substance at the surface looks like alkane, it would thus be expected that the surface tension of the surface-active substance's solution would decrease from 72 mN/m (surface tension of pure water) to alkane-like surface tension (close to 25 mN/m).

The orientation of the surface molecule at the interface will be dependent on the system. This is shown as follows:

Air–water: polar part toward water and hydrocarbon part toward air
Oil–water: polar part toward water and hydrocarbon part toward oil
Solid–water: polar part toward water and hydrocarbon part toward solid

3.2.1.1 Aqueous Solution of Surfactants
The solution properties of various surfactants in water are very unique and complex in many aspects compared to such solutes as NaCl or ethanol. However, in the following text, a very simplified but useful and practical description will be given (for details, the reader is advised to consult the relevant references [Birdi, 2002; Tanford, 1980]). The solubility of charged and noncharged surfactants are very different, especially regarding the effect of temperature and salts (such as NaCl). These characteristics are important in the application of these substances in diverse systems. For instance, one cannot use the same soap with seawater, the main reason being that salts (such as Ca and Mg) found in seawater affect the foaming and solubility characteristics of major surface-active substances. For similar reasons, one cannot use a nonionic detergent for shampoos (only amnionic detergents are used). Therefore, tailor-made surface-active agents have been devised by the soap and detergent industry to meet these specific demands.

3.2.1.1.1 Solubility Characteristics of Surfactants in
Water (Dependence on Temperature)
The solubility characteristics of surfactants (in water) is one of the most studied phenomena. Even though the molecular structures of surfactants are rather simple, their solubility in water is rather complex as compared to other amphiphiles such as long-chain alcohols, etc., in that it is dependent on the alkyl group. This is easily seen since the alkyl groups will behave mostly as alkanes. The hydrophobic alkyl part exhibits solubility in water, which has been related to a surface tension model of the cavity (see Appendix B). However, it is found additionally that the solubility

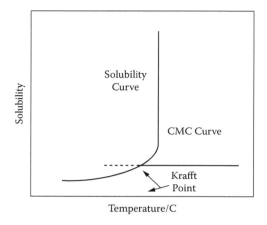

FIGURE 3.6 Solubility (Krafft point: KP) of ionic (anionic or cationic) surfactants in water (as a function of temperature).

of surfactants is also dependent on the charge present on the polar group. Ionic surfactants exhibit different solubility characteristics compared to nonionic surfactants, with regard to dependence on temperature.

Ionic surfactants: The solubility of all ionic surfactants (both anionics, that is, negatively charged, and cationics, that is, positively charged) is low at low temperature, but at a specific temperature it suddenly increases (Figure 3.6). For instance, the solubility of SDS at 15°C is about 2 g/L. This temperature is called *Krafft point* (KP), and can be obtained by cooling an anionic surfactant solution (ca. 0.5 M) from a high to a lower temperature until cloudiness appears sharply. KP is not very sharp in the case of impure surfactants that are generally found in the soap and detergent industry.

In fact, solubility near the KP is almost equal to the critical micelle concentration (CMC). The magnitude of KP is dependent on the chain length of the alkyl chain (Figure 3.7).

The linear dependence of KP on alkyl chain length is very clear. The KP for C12 sulfate is 21°C, and it is 34°C for C14 sulfate. It may be concluded that KP increases by approximately 10°C per CH_2 group. It is also interesting to note that the KP of C8 is found from extrapolation to be −3.5°C. In fact, for C8 it is not possible to measure this KP from experiments.

Since no *micelles* can be formed below the KP, it is important that one keep this information in mind when using any anionic detergent. Therefore, the effect of various parameters on KP needs to be considered in the case of ionic surfactants. The following are some of these:

Alkyl chain length (KP increases with alkyl chain length).
KP decreases if a lower-chain surfactant is mixed with a longer-chain surfactant.

Nonionic surfactants: The solubility of nonionic surfactants in water is completely different from that of charged surfactants (especially concerning the effect of

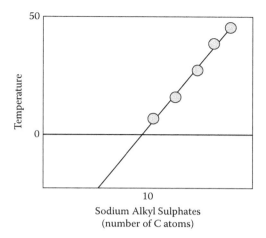

FIGURE 3.7 Variation of KP with chain length of sodium alkyl sulfates.

temperature). It is high at low temperature, but decreases abruptly at a specific temperature called the *cloud point* (CP) (Figure 3.8). (This term refers to the temperature at which the solution becomes cloudy.) This means that nonionic detergents will not be suitable if used *above* the CP. The solubility of such detergent molecules in water is due to hydrogen bond formation between the hydroxyl (-OH) and ethoxy groups ($-CH_2CH_2O-$) and water molecules. At high temperatures, the degree of hydrogen bonding gets weaker (due to high molecular vibrations), and thus nonionic detergents become insoluble at CP. The solution separates into two phases: a rich water phase and one of low concentration of nonionic surfactant. The rich nonionic detergent phase is found to have low water content. Experiments have shown that there are roughly four molecules of bound water per ethylene oxide group ($-CH_2CH_2O-$) (Birdi, 2008).

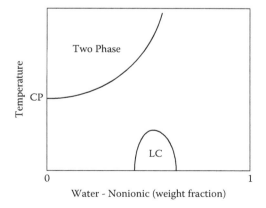

FIGURE 3.8 Solubility of a nonionic surfactant in water (cloud point: CP) (schematic) dependent on temperature.

It thus must be noted that, when a surfactant is needed for any application, its solubility characteristics, besides other properties, which should conform to the experimental conditions, should be considered. Therefore, a detergent manufacturer, for example, must be in constant collaboration with the washing industry. The surfactants available in the industry are characterized by their area of application. For instance, anionic surfactants are used in shampoos and washing, while cationics are used in hair conditioners. Hair surface has a negative (–) charge, and thus cationics strongly adsorb there, leaving a smooth surface. The charged end is oriented toward the hair surface, and the alkyl group is pointing away (as depicted here).

Cationic detergent + hair: -------alkyl group ... polar group(+)hair(–)

3.2.1.2 Micelle Formation (in Aqueous Media)

The surfactant aqueous solutions manifest two major forces that determine their behavior. The alkyl part, being hydrophobic, tends to separate out as a distinct phase, whereas the polar part tends to stay in solution. The difference between these two opposing forces thus determines the properties of the solution. The factors to be considered are the following:

a. Interaction of the alky group and water
b. Interaction of the alkyl hydrocarbon groups among themselves
c. Solvation (through hydrogen bonding and hydration with water) of the polar groups
d. Interactions between the solvated polar groups

Below CMC, the detergent molecules are present as single monomers. Above CMC, they are present as monomers, C_{mono}, in equilibrium with micelles, C_{mice}. The micelle with aggregation number, N_{ag}, is formed from monomers:

$$N_{ag} \text{ monomer} = \text{micelle} \qquad (3.1)$$

N_{ag} monomers, which were surrounded by water, aggregate together above CMC and form a micelle. In this process, the alkyl chains have transferred themselves from the water phase to an alkane-like micelle interior. This occurs because the alkyl part has a lower energy in micelle than in the water phase:

A. Water phase: Alkyl chain in water, surrounded by water
B. Micelle: Alkyl chain in contact with neighboring alkyl chains

Thus, in the case B, the repulsion between the alkyl chain and water has been removed. Instead, the alkyl–alkyl attraction (B) is the driving force for micelle formation. The surfactant molecule forms a micellar aggregate at a concentration higher than CMC because it moves from the water phase to the micelle phase (lower energy). The micelle reaches an equilibrium after a certain number of monomers have formed a micelle. This means that there are both attractive and opposing forces involved in this process; otherwise, very large aggregates would result. Thus, the standard free energy of a micelle formation, ΔG°_{mice}, could be written as:

$$\Delta G^\circ_{mice} = \text{attractive forces} + \text{opposing forces} \qquad (3.2)$$

The attractive forces are associated with the hydrophobic interactions between the alkyl part (alkyl–alkyl chain attraction) of the surfactant molecule, $\Delta G_{hydrphobic}$. The opposing forces arise from the polar part (charge–charge repulsion, polar group hydration), ΔG_{polar}, and are of opposite signs. The attractive forces would lead to larger aggregates. The opposing forces would hinder the aggregation. A micelle with a definite aggregation number is one in which the value of ΔG°_{mice} is 0. Hence, we can write ΔG°_{mice}

$$\Delta G^\circ_{mice} = \Delta G_{hydrophobic} + \Delta G_{polar} \qquad (3.3)$$

The standard free energy of micelle formation will be

$$\Delta G^\circ_{mice} = \mu^\circ_{mice} - \mu^\circ_{mono}$$
$$= RT \ln(C_{mice}/C_{mono}) \qquad (3.4)$$

At CMC, one may neglect C_{mice}, which leads to

$$\Delta G^\circ_{mice} = RT \ln(CMC) \qquad (3.5)$$

This relation holds for nonionic surfactants, but will be modified in the case of ionic surfactants (as shown in the following text). This equilibrium shows that, if we dilute the system, micelles will break down to monomers to achieve equilibrium. This is a simple equilibrium for a nonionic surfactant. In the case of ionic surfactants, there will be charged species.

In the case of ionic surfactants, such as SDS, the micelle with aggregation number N_{SD-} will consist of counterions, C_{S+}:

$$N_{SD-}\text{ionic surfactant monomers} + C_{S+}\text{ counter} = \text{micelle with charge}$$
$$(N_{SD-} - C_{S+}) \qquad (3.6)$$

Since N will be larger than S^+, all anionic surfactants are negatively charged. Similarly, cationic micelles will be positively charged. For instance, the cationic surfactant cetyltrimethyl ammonium bromide (CTAB), we have following equilibrium in micellar solutions:

$$CTAB\text{—dissociates into }CTA^+\text{ and }Br^-\text{ ions}$$

The micelle with N_{CTA+} monomers will have C_{Br-} counterions. The positive charge of the micelle will be the sum of positive and negative ions ($N_{CTA+} + C_{Br-}$). The actual concentration of each species will vary with the total detergent concentration, as in the case of SDS (Figure 3.9).

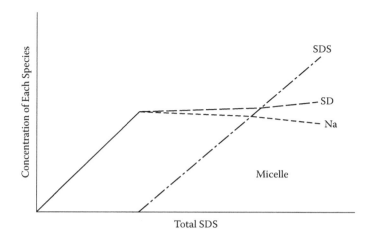

FIGURE 3.9 Variation of concentration of different ionic species for SDS solutions (Na^+, SD^-, $SDS_{micelle}$).

Below CMC, the SDS molecules in water are found to dissociate into SD^- and Na^+ ions. Conductivity measurements show that

a. SDS behaves as a strong salt, and SD^- and Na+ ions are formed (same as one observes for NaCl).
b. A break on the plot is observed at SDS concentration equal to the CMC. This clearly shows that the number of ions decreases with concentration.

The latter indicates that some ions (in the present case, cations, Na^+) are partially bound to the SDS-micelles, which results in a change in the slope of the conductivity of the solution. Similar behavior is observed for other ionic detergents such as cationic (CTAB) surfactants.

At CMC, micelles (aggregates of SD^- with some counterions, such as Na^+) are formed, and some Na^+ ions are bound to these, which is also observed from conductivity data. In fact, these data analyses have shown that approximately 70% Na^+ ions are bound to SD^- ions in the micelle. The surface charge was estimated from conductivity measurements (Birdi, 2002). Therefore, the concentration of Na^+ will be higher than SD^- ions after CMC. A large number of reports are found in the literature, in which the transition from the monomer phase (before CMC) to the micellar phase (after CMC) have been analyzed.

This is also true in the case of cationic surfactants. Thus, in the case of CTAB solutions, there are CTA^+ and Br^- ions below CMC, and above CMC, there are, additionally, CTAB micelles. In these systems, the counterion is Br^-.

3.2.1.2.1 Analyses of CMC

CMC will be dependent on different factors, including both the alkyl part and polar part. Further, the interaction of the detergent with the solvent will also have an effect on the CMC.

Effect of alkyl chain length: It has been found that CMC decreases with increasing alkyl chain length.

The following relation has been found for Na-alkyl sulfate detergents:

$$\ln(\text{CMC}) = k_1 - k_2 (C_{\text{alkyl}}) \tag{3.7}$$

where k_1 and k_2 are constants, and C_{alkyl} is the number of carbon atoms in the alkyl chain.

CMC will change if the additive has an effect on the monomer–micelle equilibrium, and also if the additive changes detergent solubility. The CMC of all ionic surfactants will decrease if coions are added. However, nonionic surfactants show very little change in CMC on the addition of salts, which is to be expected from theoretical considerations. The change of CMC with NaCl for SDS is as follows (Figure 3.10):

NaCl (mol/L)	CMC (mol/L)	g/L	Nagg
0	0.008	2.3	80
0.01	0.005	1.5	90
0.03	0.003	0.09	100
0.05	0.0023	0.08	104
0.1	0.0015	0.05	110
0.2	0.001	0.02	120
0.4	0.0006	0.015	125

The radius of the spherical micelle is reported to be 20 Å, which increases to 23 Å for the nonspherical micelle.

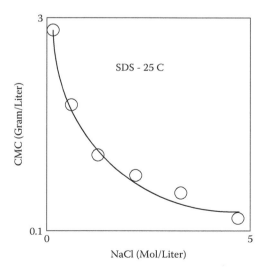

FIGURE 3.10 Variation of CMC with added NaCl for micelles.

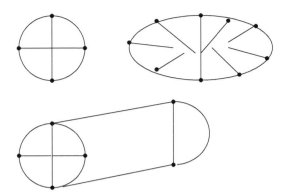

FIGURE 3.11 Different types of micellar aggregates: (a) spherical; (b) disc-like; (c) cylindrical; (d) lamellar; (e) vesicular (schematic).

Experiments have shown that, in most cases, such as for SDS, the initially spherical-shaped micelles may be influenced to grow under into larger aggregates (disc-like, cylindrical, lamellar, vesicular) (Figure 3.11).

It is noteworthy how CMC changes with even very small additions of NaCl.

It has been found that, in general, the change in CMC with the addition of ions follows the relation

$$\ln(CMC) = \text{Constant}_1 - \text{Constant}_2 \, (\ln(CMC + C_{ion})) \tag{3.8}$$

The quantity, Constant_2, was found to be related to the *degree of micelle charge*. Its magnitude varied from 0.6 to 0.7, which means that micelles have a 30% charge.

The following data were reported for the CMC of cationic surfactants, which decreased on addition of KBr as follows:

DTAB (dodecyltrimethylammonium bromide) and TrTAB:

$$\ln(CMC) = -6.85 - 0.64 \ln(CMC + C_{KBr}) \tag{3.9}$$

TrTAB:

$$\ln(CMC) = -8.10 - 0.65 \ln(CMC + C_{KBr}) \tag{3.10}$$

TTAB:

$$\ln(CMC) = -9.43 - 0.68 \ln(CMC + C_{KBr}) \tag{3.11}$$

It is noticed that the slope increases with increase in alkyl chain length. A similar relationship has been reported for the Na-alkyl sulfate homologous series.

The CMC data for soaps is found to give the following dependence on alkyl chain length:

Soap	CMC (mol/L)25°C
C7COOK	0.4
C9COOK	0.1
C11OOK	0.025
C7COOCs	0.4

These data show that CMC decreases by a factor of 4 *for each increase in chain length* by -CH$_2$CH$_2$-.

3.3 GIBBS ADSORPTION EQUATION IN SOLUTIONS

A pure liquid (such as water), when shaken, does not produce any foam. This merely indicates that the surface layer consists of pure liquid (and there is absence of any minor surface-active impurities). However, if a very small amount of surface-active agent is added (soap or detergent of about millimolar concentration or about parts per million by weight is added) and the solution shaken, foam is produced at the surface of the solution. This indicates that the surface-active agent has *accumulated* at the surface (meaning that the concentration of the surface-active agent is much higher at the surface than in the bulk phase; in some cases, many thousand times) and thus forms a thin-liquid film (TLF) that constitutes the bubble. In fact, one can use bubble or foam formation as a useful criterion for the purity of the system. It can be generally observed at the shores of lakes or ocean that foam bubbles are formed under different conditions. If water in these sites is polluted, then very stable foams are observed.

If one adds an inorganic salt, such as NaCl, instead of detergent, then no foam is formed. Foam formation indicates that the surface-active agent adsorbs at the surface, and forms a TLF (consisting of two layers of amphiphile molecules and some water). This has led to many theoretical analyses of surfactant concentration (in the bulk phase) and surface tension (consequent on the presence of surfactant molecules at the surface). The thermodynamics of surface adsorption has been extensively described by the Gibbs adsorption theory (Chattoraj and Birdi, 1984).

Further, the Gibbs adsorption theory has also been used for systems other than solutions (such as solid–liquid or liquid$_1$–liquid$_2$, adsorption of solute on polymers, etc.). In fact, the Gibbs adsorption theory will be applicable to any system in which adsorption takes place at an interface.

3.3.1 GIBBS ADSORPTION THEORY AT LIQUID INTERFACES

The surface tension γ of water changes with the addition of organic or inorganic solutes at constant temperature and pressure (Defay et al., 1966; Chattoraj and Birdi, 1984; Birdi, 1989, 1997). The extent of surface tension change and the sign of the change are determined by the molecules involved (see Figure. 3.1).

The γ values of aqueous solutions generally increase with different electrolyte concentrations. The magnitude of γ of aqueous solutions containing organic solutes invariably decreases. As mentioned earlier, the surface of a liquid is where the density of the liquid changes to that of a gas by a factor 1000 (Chapter 1, Figure 1.1).

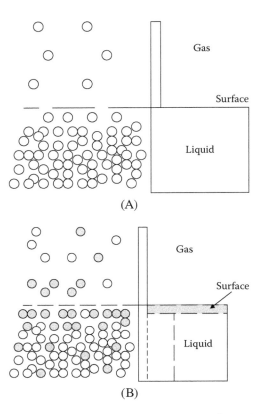

FIGURE 3.12 Surface composition: (A) of pure water, (B) of an ethanol–water solution (shaded = ethanol).

Now let us look at what happens to the surface composition when ethanol is added to water (Figure 3.12 [A,B]).

The reason why ethanol concentration in the vapor phase is higher than that of water is its lower boiling point. Next, let us consider the situation when a detergent is added to water, whereby the surface tension is lowered appreciably (Figure 3.13).

The schematic concentration profile of detergent molecules is such that the concentration is homogenous up to the surface. At the surface, there is almost only detergent molecules plus the necessary number of water molecules (which are in a bound state to the detergent molecules). The surface thus shows very low surface tension (ca. 30 mN/m). The surface concentration profile of the detergent is not easily determined by any direct method. In Figure 3.13, it is shown as a rectangle for convenience, but one may also imagine other forms of profiles, such as curved.

It can be observed that surface tension decreases due to ethanol. This suggests that there are more ethanol molecules at the surface than in the bulk phase, which is also seen in a cognac glass. The ethanol vapors are observed to condense on the edge of the glass, showing that the concentration at the surface of the solution is very high.

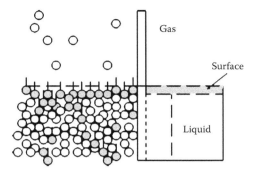

FIGURE 3.13   Concentration of detergent (shaded with tail) in solution and at the surface. The shaded area at the surface is the excess concentration due to accumulation.

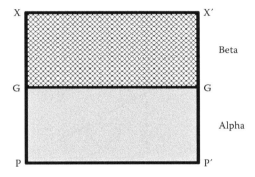

FIGURE 3.14   Liquid column in a real system with α-phase and β-phase.

To analyze these data, the well-known Gibbs adsorption equation (Chattoraj and Birdi, 1984; Birdi, 1989) has to be used. A liquid column containing i number of components is shown in Figure 3.14, according to the Gibbs treatment of two bulk phases, that is, α and β, separated by the interfacial region AA'BB'.

Gibbs considered that this interfacial region is inhomogeneous and difficult to define, and he therefore also considered a more simplified case in which the interfacial region is assumed to be a mathematical plane GG' (Figure 3.14).

In the actual system (Figure 3.15), the bulk composition of the i-th component in α-phase and β-phase are $c_{i\alpha}$ and $c_{i\beta}$, respectively.

However, in the idealized system, the chemical compositions of the α and β phases are imagined to remain unchanged right up to the dividing surface so that their concentrations in the two imaginary phases are also $c_{i\alpha}$ and $c_{i\beta}$, respectively.

If $n_{i\alpha}$ and $n_{i\beta}$ denote the total moles of the i-th component in the two phases of the idealized system, then the Gibbs surface excess Γ_{ni} of the i-th component can be defined as

$$n_i^x = n_i^t - n_{i\alpha} - n_{i\beta} \tag{3.12}$$

where n_i^t is the total moles of the i-th component in the real system.

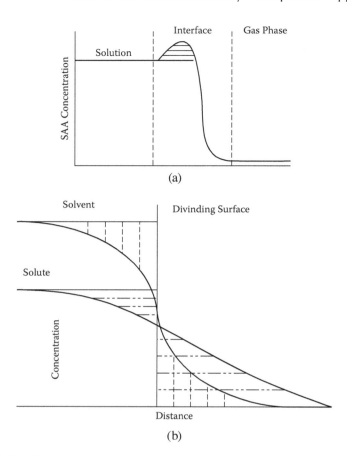

FIGURE 3.15 Liquid composition column in an ideal system (schematic).

In an exactly similar manner, the respective surface excess internal energy E_x and entropy S_x can be defined by the following mathematical relationships (Chattoraj and Birdi, 1984; Birdi, 1989):

$$E^x = E^t - E^\alpha - E^\beta \qquad (3.13)$$

$$S^x = S^t - S^\alpha - S^\beta \qquad (3.14)$$

Here, E^t and S^t are the total energy and entropy, respectively, of the system as a whole for the actual liquid system. The energy and entropy terms for the α and β phases are denoted by the respective superscripts. The excess (x) quantities thus refer to the surface molecules in the adsorbed state.

The real and idealized systems are open so that following equation can be written:

$$dE^t = T\, dS^t - (p\, dV + p'\, dV' - \gamma\, dA) + \mu_1\, dn_{1t} +$$
$$\mu_2\, 2\, dn_{2t} + \ldots + \mu_i\, dn_i^t \qquad (3.15)$$

where V^α and V^β are the actual volumes of each bulk phase, and p and p′ are the respective pressures. Since the volume of the interfacial region was considered to be negligible, $V^t = V^\alpha + V^\beta$. Further, if the surface is almost planar, then $p^\alpha = p'^\beta$, and $(p\, dV^\alpha + p_\beta\, dV^\beta) = p\, dV^t$.

The changes in the internal energy for idealized phases α and β may similarly be expressed as follows:

$$dE^\alpha = T\, dS^\alpha - p\, dV^\alpha + \mu_i\, dn_i + \dots + \mu_i,\, dn_{i\alpha} \tag{3.16}$$

and

$$dE^\beta = T\, dS^\beta - p\, dV^\beta + \mu_i\, dn_i + \dots + \mu_{i,\beta}\, dn_{i,\beta} \tag{3.17}$$

In the real system, the contribution due to change in the surface energy, $\gamma\, dA$, is included as an additional work. Such a contribution is absent in the idealized system containing only two bulk phases without the existence of any physical interface.

By subtracting Equations 3.16 and 3.17 from Equation 3.15, the following relationship is obtained:

$$d(E^t - E^\alpha - E^\beta) = T\, d(S^t - S^\alpha - S^\beta) + \gamma\, dA + \mu_i\, d(nt - n1 - n1) + \dots +$$
$$\mu_i\, d(n_{i,t} - n_i - n_{i\,\beta}) \tag{3.18}$$

or

$$d\, E^x = T\, dS^x + \gamma\, dA + \mu_1\, dn_1 x + \dots + \mu_i\, dn_{ix} \tag{3.19}$$

This equation, on integration at constant T, γ, and μ_i, etc., gives

$$E^x = T\, S^x + \gamma\, A + \mu_1\, n_1\, ix + \dots + \mu_1\, n_{ix} \tag{3.20}$$

This relationship may be differentiated in general to give

$$dE^x = T\, dS^x + \gamma\, dA + \Sigma_i\, (\mu_i\, dn_i) + \Sigma_i\, n_i\, d\mu_i + A\, d\gamma + S\, dT \tag{3.21}$$

Combining Equations 3.21 and 3.20 gives

$$-A\, d\gamma = S^x\, dT + \Sigma_i\, n_{ix}\, d\mu_i \tag{3.22}$$

Let $S^{s,x}$ and Γ_{ix} denote the surface excess entropy and moles of the i-th component per surface area, respectively. This gives

$$S^{s,x} = S^x/A$$

$$\Gamma_{ix} = n_{ix}/A \tag{3.23}$$

and

$$-d\,\gamma = S^{s,x}\,dT + \Gamma_{x,1}\,d\mu_1 + \Gamma_{x,2}\,d\mu_2 + \ldots + \Gamma_{ix}\,d\mu_i \qquad (3.24)$$

This equation is similar to the Gibbs–Duhem equations for the bulk liquid system.

To make this relationship more meaningful, Gibbs further pointed out that the position of the plane may be shifted parallel to GG' along the x-direction and fixed in a particular location, when (n_{it}) becomes equal to $(n_1 + n_i)$. Under this condition, n_{ix} (n_1 by convention) becomes zero. The relation in Equation 3.24 can be rewritten as

$$-d\,\gamma = S_{s,1}\,dT + \Gamma_2\,d\mu_2 + \ldots + \Gamma_i\,d\mu_i \qquad (3.25)$$

$$= S_{s,1}\,dT + 3\,\Gamma_i\,d\mu_i \qquad (3.26)$$

At constant T and p, for a two-component system (say, water(1) + alcohol(2)), we thus obtain the classical Gibbs adsorption equation as

$$\Gamma_2 = -\,(d\,\gamma/d\mu_2)_{T,p} \qquad (3.27)$$

The chemical potential μ_2 is related to the activity of alcohol by the equation

$$\mu_2 = \mu_2^{\circ} + R\,T\,\ln(a_2) \qquad (3.28)$$

If the activity coefficient can be assumed to be equal to unity, then

$$\mu_2 = \mu^{\circ}_2 + R\,T\,\ln(C_2) \qquad (3.29)$$

where C_2 is the bulk concentration of solute 2. Then, the Gibbs adsorption can be written as

$$\Gamma_2 = -1/RT(d\,\gamma/d\,\ln(C_2))$$
$$= -C_2/RT\,(d\,\gamma/dC_2) \qquad (3.30)$$

This shows that the *surface excess* quantity on the left-hand side is proportional to the change in surface tension with concentration of the solute (d γ/d($\ln(C_{surfaceactivesubstance})$). A plot of ln (C_2) versus γ gives a slope equal to

$$\mathbf{\Gamma 2\ (RT)} \qquad (3.31)$$

from which the value of Γ_2 (moles/area) can be estimated.

This shows that all surface-active substances will always have a higher concentration at the surface than in the bulk of the solution.

This relation has been verified using radioactive tracers. Further, as will be shown later under spread monolayers, there is very convincing support to this relation and the magnitudes of Γ for various systems.

The surface tension of water (72 mN/m, at 25°C) decreases to 63 mN/m in a solution of SDS of concentration 1.7 mmol/L. The large decrease in surface tension suggests that SDS molecules are concentrated at the surface, as otherwise there should be very little change in the surface tension. This means that the concentration of SDS at the surface is much higher than in the bulk. The molar ratio of SDS:water in the bulk is 0.002:55.5. At the surface, the ratio will be expected to be of a completely different value, as found from the value of Γ (the ratio is 1000:1). This is also obvious when considering that foam bubbles form on solutions with very low surface-active agent concentrations. The foam bubble consists of a bilayer of surface-active agent with water inside. In fact, it is easy to consider the state of surfactant solutions in terms of molecular ratios. If we use SDS, for example, since it is a strong electrolyte, it can be considered to dissociate completely:

$$C_{12}\,H_{25}\,SO_4\,Na \rightarrow C_{12}H_{25}SO_4- + Na^+$$

$$SDS \rightarrow DS^- + S^{+\prime} \tag{3.32}$$

The appropriate form of the Gibbs equation will be

$$-d\,\gamma = \Gamma_{DS} - d\mu_{DS} - + \Gamma_{S+}\,d\mu_{s+} \tag{3.33}$$

where the surface excess, Γ_i, terms for each species in the solution (e.g., DS– and S+) are included. This equation relates the observed change in surface tension to the changes in the chemical potential of the respective solutes (here DS– and S+). The chemical potential terms can be expanded in an analogous way to Equation 3.17, and then we obtain

$$-d\,\gamma = RT\,[\Gamma_{DS} - d(\ln C_{DS-}) + \Gamma_S + d(\ln C_{s+})] \tag{3.34}$$

$$= RT\,(\Gamma_{DS} - dC_{DS}/C_{DS-} + \Gamma_S + dCs+/dCs+) \tag{3.34a}$$

In order to simplify this equation, we must assume that electrical neutrality is maintained in the interface; then we may write

$$\Gamma_{SDS} = \Gamma_{DS} - = \Gamma_S+ \tag{3.35}$$

and

$$C_{SDS} = C_{DS-} = C_{S+} \tag{3.36}$$

which, on substitution in Equation 3.34, gives

$$-d\,\gamma = 2\,RT\,\Gamma_{SDS}\,d(\ln C_{SDS}) \tag{3.37}$$

$$= (2RT/C_{SDS})\,\Gamma_{SDS}\,dC_{SDS} \tag{3.38}$$

and one obtains

$$\Gamma_{SDS} = -1/2(R\ T)\ d\ \gamma/(d\ \ln\ C_{SDS}) \qquad (3.39)$$

In the case where the ion strength is kept constant, that is, in the presence of added NaCl, then the equation becomes

$$\Gamma_{SDS} = -1/(R\ T)\ d\ \gamma/(d\ \ln\ C_{SDS}) \qquad (3.40)$$

Comparing Equation 3.39 with Equation 3.40, it will be seen that they differ by a factor of 2, and that the appropriate form will need to be used in the experimental test of the Gibbs equation. It is also quite clear that any partial ionization would lead to considerable difficulty in applying the Gibbs equation. Finally, while on this topic, if SDS were investigated in a solution using a large excess of sodium ions, produced by the addition, say, of NaCl, then the sodium ion term in Equation 3.39 will vanish and will arrive back at an equation equivalent to Equation 3.40.

The adsorption of detergents at the surface of the solution can be estimated by the Gibbs equation. It is convenient to plot γ versus the log $(C_{detergent})$.

From the γ versus $C_{alkyl\ sulfate}$ data, the following data is obtained from the Gibbs equation:

Concentration mol/L	Γ_{Salkyl} Sulfate 10^{-12}mol/cm^2	A (area/molecule)
NaC$_{10}$ sulfate (0.03 mol/L)	3.3	50 Å^2
NaC$_{12}$ sulfate (0.008 mol/L)	3.4	50 Å^2
NaC$_{14}$ sulfate (0.002 mol/L)	3.3	50 Å^2

From the plots of γ versus concentration, the slope is related to the surface excess, $\Gamma_{Salkyl\ sulfate}$. The area/molecule values indicate that the molecules are aligned vertically on the surface, irrespective of the alkyl chain length. If the molecules were oriented flat, then the value of area/molecule would be much larger (approximately 100 Å^2). Further, the fact that the alkyl chain length has no effect on the area also proves this assumption. These conclusions have been verified from spread monolayer studies. Further, it is also found that the polar group, that is, $-SO_4^-$, would occupy something like 50 Å^2. Later, it will be shown that other studies confirm that the area per molecule is approximately 50 Å^2.

The Gibbs adsorption equation is a relation about the solvent and a solute (or many solutes). The solute is present either as excess (if there is an excess surface concentration) if the solute decreases the γ, or as a deficient solute concentration (if the surface tension is increased by the addition of the solute).

It can be further explained that, if we consider the system, water, to which a surfactant such as SDS is added, the molecules at the surface will change as follows:

pure water (**w**) surface:

wwwwwwwwwwwwwwwwwwwwwwwwww
wwwwwwwwwwwwwwwwwwwwwwwwww
wwwwwwwwwwwwwwwwwwwwwwwwww

water (w) plus SDS(s) (2 g/L):

sssWsssssssssWsssssssssssssssssWssssWWsssssssssssWs
ssssWsssWWWWWsWWWsssssWWWWWWWWWWWW
ssssWsssWWWWWsWWWsssssWWWWWWWWWWWW

This shows that the surface tension of pure water of 72 mN/m decreases to 30 mN/m by the addition of 2 g/L of SDS. Thus, the surface of the SDS solution is mostly a monolayer of SDS plus some bound water. The ratio of water:SDS in the system is roughly as follows:

In the bulk phase: 55 mol water: 8 mmol SDS
At the surface: roughly 100 mol SDS: 1 mol water

This description is in accord with the decrease in γ of the system.

Investigations have shown that, if one carefully sucked a small amount of the surface solution of a surfactant, then one can estimate the magnitude of Γ. The concentration of the surface-active substance was found to be 8 μmol/mL. The concentration in the bulk phase was 4 μmol/L. The data show that the surface excess is 8 μmol/mL $-$ 4 μmol/mL $=$ 4 μmol/mL. Further, this indicates that, when there is 8 μmol/L in the bulk of the solution, the SDS molecules completely cover the surface. The consequence of this is that, at a concentration higher than 8 μmol/L, no more adsorption at the interface of SDS takes place. Thus, γ remains constant (almost). This means that the surface is completely covered with SDS molecules. The area-per-molecule data (as found to be 50 Å^2) indicates that the SDS molecules are oriented with the SO4$-$ groups pointing toward the water phase, while the alkyl chains are oriented away from the water phase.

This means if one, through foam bubbles, collected the foam continuously, then more and more surface-active substances will be removed. Such a method of bubble foam separation has been used to purify wastewater from surface-active substances. It is especially useful when very minute amounts of surface-active substances (dyes in the printing industry; pollutants in wastewater). The method is economical to use and is free of any chemicals or filters. In fact, if the pollutant is very expensive or poisonous, then this method can have many advantages over the other methods.

It is useful to consider the amount of SDS in a bubble of radius 1 cm. Assuming that there is almost no water in the bilayer of the bubble, the surface area of the bubble can be used to estimate the amount of SDS. We can use the following data:

Radius of bubble $=$ 1 cm
Surface area $=$ (4 Π 1^2) 2 $=$ 25 cm^2 $=$ 25 10^{16} Å^2
Area per SDS molecule $=$ 55 Å^2
Number of SDS molecules per bubble $=$ 0.5 10^{16} molecules
Amount of SDS per bubble $=$ 0.5 10^{16}/6 10^{23} g
$=$ 0.01 μg SDS

It can thus be seen that it would require 100 million bubbles to remove 1 g of SDS from the solution. Since bubbles can be easily produced at very fast rates (ca.

100–1000 bubbles per minute), this is not a big hindrance. Consequently, any kind of other surface-active substances (such as pollutants in industry) can thus be removed by foaming.

During the past decades, a few experiments have been reported, in which verification of Gibbs adsorption has been reported. One of these methods has been carried out by removing by a microtone blade the thin layer of surface of a surfactant solution. Actually, this is almost the same as the procedure of bubble extraction, that is, merely by a careful suction of the surface layer of solution. The surface excess data for a solution of SDS were found to be acceptable. The experimental data was 1.57 10^{-18} mol cm^{-2}, while from the Gibbs adsorption equation one expected it to be 1.44 10^{-18} mol cm^{-2}.

Example: A solution of CTAB shows following data:

$$\gamma = 47 \text{ mN/m}, C_{ctab} = 0.6 \text{ mmol/L}$$

$$\gamma = 39 \text{ mN/m}, C_{ctab} = 0.96 \text{ mmol/L}$$

From these equations, one can calculate

$$d \gamma / d \log(C_{ctab}) = (47 - 39)/(\log(0.6) - \log(0.96)) = 8/(-0.47) = 17$$

From the foregoing data, the area/molecule for CTAB is found to be 90 Å^2, which is reasonable.

3.3.2 KINETIC ASPECTS OF SURFACE TENSION OF DETERGENT AQUEOUS SOLUTIONS

Without exception, information is needed on the kinetic aspects of any phenomena. In the present case, the question to be asked is how fast the surface tension of a detergent solution reaches equilibrium.

If one pours a detergent solution into a container, then the instantaneous concentration of the detergent will be uniform throughout the system; that is, it will be the same in the bulk and at the surface. Since the concentration of the surface-active substance is very low, the surface tension of the solution will be the same as that of pure water (i.e., 72 mN/m, at 25°C). This is because the surface excess at time zero is zero. However, it is found that the freshly formed surface of a detergent solution exhibits varying rates of change in surface tension with time. A solution is uniform in solute concentration until a surface is created. At the *surface*, the surface-active substance will accumulate and surface tension will decrease with time. In some cases, the rate of adsorption at the surface is very fast (less than a second), while in other cases, it may take a longer time (Figure 3.16). One finds that the freshly created aqueous solution shows surface tension of almost pure water, that is, 70 mN/m. However, surface tension starts to decrease rapidly and reaches an equilibrium value (which may be lower than 40 mN/m) after a given time. In general, this phenomenon has no consequence. However, in some cases where a fast cleaning process is involved, then one must consider the kinetic aspects. In fact, the formation of foam bubbles as one

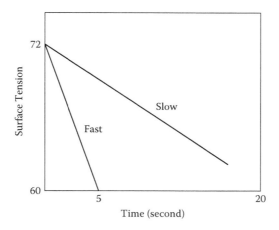

FIGURE 3.16   Kinetics of adsorption of detergent at an interface (schematic).

pours the solution is indicative of that surface adsorption is indeed very fast (as pure water does not foam on shaking!).

Especially in the case of high-molecular-weight surface-active substances (such as proteins), the period of change may be sufficiently prolonged to allow easy observation. This arises because proteins are surface active. All *proteins* behave as surface-active substances because of the presence of hydrophilic–lipophilic properties imparted from the different *polar*, such as glutamine and lysine, and *apolar*, such as alanine, valine, phenylalanine, isovaline, amino acids. Proteins have been extensively investigated as regards their polar–apolar characteristics as determined from surface activity.

Based on simple diffusion assumptions, the rate of adsorption at the surface,

$$d\Gamma/dt = (\mathbf{D}/\pi)^2 \, C_{bulk} \, t^{-2} \tag{3.41}$$

which on integration gives

$$\Gamma = 2 \, C_{bulk} \, (\mathbf{D}t/\pi)^2 \tag{3.42}$$

where $\mathbf{D}$ is the diffusion constant coefficient, and C_{bulk} is the bulk concentration of the solute. The procedure used is to apply suction at the surface, and the fresh surface is created instantaneously (Birdi, 1989). The magnitude of γ increases to that of pure water (i.e., 72 mN/m) and decreases with time as Γ increases (from the initial value of zero). This experiment actually verifies the various assumptions as made in the Gibbs adsorption equation. Experimental data shows good correlation to this equation when t is very small.

3.3.3   SOLUBILIZATION (OF ORGANIC-WATER-INSOLUBLE MOLECULES) IN MICELLES

In many everyday needs, one should be able to apply organic-water-insoluble compounds in industry and biology. It has been found that micelles (both ionic and nonionic) behave as a micro phase, where the *inner core* behaves as (liquid) alkane, while the surface area behaves as a polar phase. The inner core is also found to exhibit liquid-*alkane-like* characteristics. The inner core thus has been found to exhibit alkane-like properties while being surrounded by a water phase. In fact, micelles are *nanostructures*. What this then suggests is that surfactant solution systems can be designed in water, which can have both aqueous and alkane-like properties. This unique property is one of the main applications of surfactant micelle solutions in all kinds of systems. Further, in ionic surfactant micelles, one can additionally create *nano-reactor* systems. In the latter reactors, the counterions are designed to bring two reactants in very close proximity (Birdi, 2007).

The most useful characteristic of the micelle arises from its inner (alkyl chain) part (Figure 3.17). The inner part consists of alkyl groups that are closely packed. It is known that these clusters behave as *liquid paraffin* (Cn H_2n+2). The alkyl chains are thus not fully extended. Hence, one would expect that this inner hydrophobic part of the micelle should exhibit properties that are common to alkanes, such as ability to solubilize all kinds of water-insoluble organic compounds. The solute enters the alkyl core of the micelle and it swells. Equilibrium is reached when the ratio between moles solute:moles detergent is reached corresponding to the thermo-dynamic value.

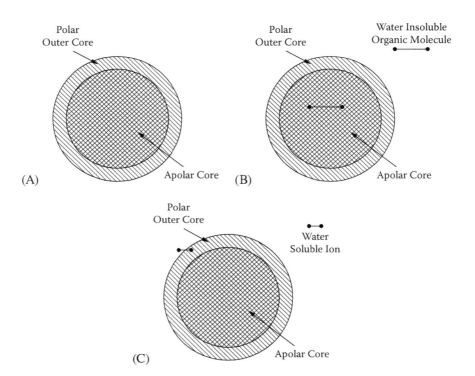

FIGURE 3.17 Micelle structure: (A) (inner part = liquid paraffin-like; outer polar part); (B) solubilization of apolar molecule; (C) binding of counterion to the polar part (schematic).

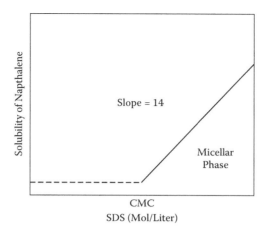

FIGURE 3.18 Solubilization of naphthalene in SDS solutions.

Size analyses of (using light scatter) some spherical micelles of SDS have indeed shown that the radius of the micelle is almost the same as the length of the SDS molecule. However, if the solute interferes with the outer polar part of the micelle, then the micelle system may change in such a way that the CMC and other properties change. This is observed in the case of the addition of dodecanol to SDS solutions. However, very small additions of solutes show very little effect on CMC. The data in Figure 3.18 show the change in the solubility of naphthalene in SDS aqueous solutions.

Below CMC the amount dissolved remains constant, which corresponds to its solubility in pure water. The slope of the plot above CMC corresponds to 14 mole SDS:1 mol naphthalene. It is seen that, at the CMC, the solubility of naphthalene abruptly increases. This is because all micelles can solubilize water-insoluble organic compounds. A more useful analysis can be carried out by considering the thermodynamics of this solubilization process.

At equilibrium, the chemical potential of a solute (naphthalene, etc.) will be given as

$$\mu_s^s = \mu_s^{aq} = \mu_s^M \tag{3.43}$$

where μ_s^s, μ_s^{aq}, and μ_s^M are the chemical potentials of the solute in the solid state, aqueous phase, and micellar phase, respectively. It must be noted that, in these micellar solutions, we will describe the system in terms of the *aqueous* phase and the *micellar* phase. The standard free energy change involved in the solubilization, ΔG°_{sol}, is given as follows:

$$\Delta G^\circ_{sol} = - R T \ln (C_{s,M}/C_{s,aq}) \tag{3.44}$$

where $C_{s,aq}$ and $C_{s,M}$ are the concentrations of the solute in the aqueous phase and in the micellar phase, respectively. The free energy change is the difference between the energy when the solute is transferred from the solid state to the micelle interior. It has been found from many systematic studies that ΔG_{sol} is dependent on the chain length of the alkyl group of the surfactant. The magnitude of ΔG_{sol} changes by -837 J (-200 cal)/mol, with the addition of the -CH2- group. In most cases, the addition of electrolytes to the solution has no effect (Birdi, 1982, 1999, 2002). The kinetics of solubilization has an effect on its applications (Birdi, 2002).

Another important aspect is that the slope in Figure 3.18 corresponds to ($1/C_{sM}$). This allows the determination of moles SDS required to solubilize one mole of the solute.

Analyses of various solutes in SDS systems showed that

Naphthalene. 14 mol SDS/mole naphthalene
Anthracene. 780 mol SDS/mole anthracene
Phenathrene. 47 mol SDS/mole phenathrene

This kind of study allows the determination (quantitatively) of the range of solubilization in any such application. These systems, when used to solubilize water-insoluble organic compounds, would require such information (in such systems as pharmaceutical, agriculture sprays, paints, etc.). The dosage of any substance is based on the amount of material per volume of a solution.

Thus, this also shows that, wherever detergents are employed, the major role (besides lower surface tension) would be the solubilization of any water-insoluble organic compounds. This process would then assist in cleaning, or washing, or any other effect. In some cases such as bile salts, the solubilization of lipids (especially lecithins) gives rise to some complicated micellar structures. Due to the formation of mixed lipid–bile salt micelles, changes are observed in CMC and aggregation number. This has major consequences for bile salts in biology.

Dietary fat consists essentially of mixed triglycerides. These fatty lipids pass through the stomach into the small intestine without much change in structure. In the small intestine, triglycerides are partly hydrolyzed by an enzyme (lipase) that leads to the formation of oil–water emulsion.

3.3.4 BILE SALT MICELLES

Bile salts are steroids with detergent properties that are used by nature to emulsify lipids in foodstuff passing through the intestine to enable fat digestion and absorption through the intestinal wall. They are secreted from the liver, stored in the gall bladder, and passed through the bile duct into the intestine when food is passing through. Bile salts in general form micelles with low aggregation numbers (ca. 10–50). However, bile micelles grow very large in size when it solubilizes lipids.

Biosynthesis represents the major metabolic fate of cholesterol, accounting for more than half of the 800 mg/day of cholesterol that the average adult uses up in metabolic processes. By comparison, steroid hormone biosynthesis consumes only about 50 mg of cholesterol per day. Much more that 400 mg of bile salts is required and secreted into the intestine per day, and this is achieved by recycling them. Most of the bile salts secreted into the upper region of the small intestine are absorbed along with the dietary lipids that they emulsified at the lower end of the small intestine. They are separated from the dietary lipids and returned to the liver for reuse. Recycling thus enables 20–30 g of bile salts to be secreted into the small intestine each day.

The most abundant of the bile salts in humans are cholate and deoxycholate, and they are normally conjugated with either glycine or taurine to give glycocholate or taurocholate, respectively. The conjugation is important in identifying the bile salt for recycling back to the liver. The cholesterol contained in bile will occasionally accrete into lumps in the gall bladder, forming gallstones.

In the absence of bile, fats become indigestible and are, instead, excreted in the feces. This causes significant problems in the parts of the intestine, as normally, virtually all fats are absorbed in the duodenum and the intestines, and bacterial flora are not adapted to processing fats past this point. The role of bile salt micelles in these biological systems is very important.

3.4 APPLICATIONS

The area of industry where surface-active agents are applied is extensive and beyond the coverage of this book. However, some main industrial applications will be described in very general terms.

3.4.1 CLEANING AND DETERGENCY

One of the most important applications of surface and colloid chemistry principles in everyday life is in the systems where cleaning and detergency is involved. These are some of the most important phenomena for humans (as regards health and welfare and technology), and it has been regarded as such for many centuries. For example, the effect of clean wings of airplanes is of utmost concern in flight security. Humans have been aware of the role of cleanliness on health and disease for many thousands of years. Many critical diseases, such as AIDS or similar infections, are found to be lesser in incidence in those areas of the world where cleanliness is highest. The term *detergency* is used for such processes as *washing clothes*, or *drycleaning*, or cleaning. The substances used are designated as detergents (Zoller, 2008). In all these processes, the object is to remove dirt from fabrics or solid surfaces (floors or walls or other surfaces of all kinds).

The shampoo is used to clean hair. Hair consists of portentous material and thus requires different kind of detergents than when washing clothes or a car. A shampoo should not interact strongly with the hair but should remove dust particles or other material. Another important requirement is that the ingredients in the shampoo should not damage or irritate the eye or skin with which it may come in contact. In fact, all shampoos are tested for eye and skin irritation before marketing. In some cases, by increasing the viscosity, one achieves a great deal of protection. For example, a surfactant solution (ca. 20%), alkyl sulfate with two EO, gives very high viscosity if a small amount of salt is added.

In the case of drycleaning, the aqueous medium is replaced by a nonaqueous medium (such as tetrachloroethylene). Tetrachloroethylene can dissolve grease from materials that should not be treated by water. Since there is also polar dirt (such as sugars, minerals, etc.), a small amount of water needs to be added. This leads to the

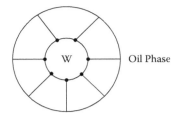

FIGURE 3.19 Structure of the inverted micelle.

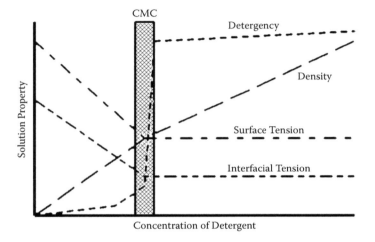

FIGURE 3.20 Change in detergent solution properties as related to CMC (schematic).

formation of the so-called *inverted micelles* (Figure 3.19). In *inverted micelles*, polar dirt is solubilized inside the water phase.

Detergents used for rug cleaning are tailor made. In the case of the process where detergent solution is applied in a highly foaming state, the principle is as follows. As the dirt is removed by the detergent solution and is present in the foam, the latter dries and can be vacuum cleaned. This requires that the detergent is not hygroscopic and thus easily removed. Appropriate salts (such as Li) of alkyl sulfates, which are very weakly hygroscopic, can be used.

In general, in all such systems, an enhanced detergency is observed when the concentration of detergent is greater than its CMC (Figure 3.20).

4 Spread Lipid Films on Liquid Surfaces (Langmuir–Blodgett Films)

4.1 INTRODUCTION

Ancient Egyptians are known to have spread very small amounts of olive oil over the water while their ships were sailing into harbors. Because this oil was known to appreciably calm the waves, navigation into the harbors was made easier.

It was found centuries later that some lipid-like substances (almost insoluble in water) formed *self-assembly monolayers* (SAMs) on the surface of water (Gaines, 1966; Adamson and Gast, 1997; Birdi, 1989, 1999, 2002; Chattoraj and Birdi, 1984). A few decades ago, experiments showed that monomolecular films of lipids could be studied by using rather simple experimental methods (Figure 4.1).

When a surface-active agent (such as surfactant or a soap) is dissolved in water and adsorbs preferentially at the surface (surface excess, Γ), it shows that the concentration of the surface-active agent may be as high as 1000 times more than in bulk. The decrease in surface tension indicates this and also suggests that only a monolayer is present at the surface. For example, in a solution of sodium dodecyl sulfate (SDS) of concentration 0.008 mol/L, the surface is completely covered with SDS molecules. In this chapter, systems consisting of lipids presenting as monolayers on water will be analyzed. In these, almost all the substance applied to the surface (in the range of few micrograms) is supposed to be present at the interface. This means that one knows quantitatively the magnitude of surface concentration (same as the surface excess, Γ).

Thin organic films of thickness of 20 Å (greater than 2 nm) or more are now found to be very useful structures. In 1774, Benjamin Franklin reported the effect of very small amounts of oil on the surface of water. However, in 1920, Langmuir was awarded the Nobel Prize for studying these monomolecular films in an apparatus (Figure 4.1).

If one places a very small amount of a lipid on the surface of water, it may affect surface tension in different ways. It may not show any effect (such as in the case of cholesterol), or it may show a drastic decrease in surface tension (such as in the case of stearic acid or tetra-decanol). An amphiphile molecule will adsorb at the

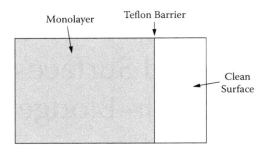

FIGURE 4.1 Monolayer film balance: barrier and surface pressure.

air–water or oil–water interface, with its alkyl group pointing away from the water phase. The alkyl group is at a lower energy state when pointing toward air than when surrounded by water molecules. The molecular properties and applications are therefore essential for investigation. The system can be easily considered as a *two-dimensional* analog to the classical *three-dimensional* systems (as indicated by the gas theory). Later experiments showed that these *monomolecular* films were relevant to many more complicated systems.

It is thus seen that the Π of a monolayer is the lowering of surface tension due to the presence of monomolecular film. This arises from the orientation of the amphiphile molecules at the air–water or oil–water interface, where the polar group would be oriented towards the water phase, while the nonpolar part (hydrocarbon) would be oriented away from the aqueous phase. This orientation produces a system with minimum free energy.

Later it was observed that, if a clean smooth solid is dipped through a monolayer, in most cases a single layer of lipid will be deposited on the solid. This film was called the *Langmuir–Blodgett* (LB) film (Figure 4.2). Further, if this process is repeated, multilayers could be deposited. This LB film technique has found many applications in the electronics industry. Further, the presence of only one layer of lipid changes the surface properties (such as the contact angle wetting, friction, light reflection, charge, adhesion, etc.) of the solid.

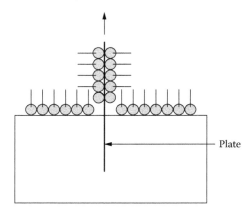

FIGURE 4.2 LB film formation (see text for details).

4.1.1 SURFACE FILM BALANCE

Since Langmuir reported monolayer studies, a great many instruments have been designed about this method. The clean surface of water shows no change in surface tension if a barrier is moved across it. However, if a surface-active agent is present, then the latter molecules will be compressed, and this will give rise to a decrease in surface tension.

The aim in these systems is to study the properties of a monomolecular thin-lipid film spread on the surface of water. Lately, it has been shown that such lipid films are useful membrane models for biological membrane structure and function studies. It is thus obvious that much careful arrangement is required in order to achieve this goal, with a high degree of cleanliness required so that contaminants do not interfere with the data. Modern methods allow measurement of monolayer properties. It is important to note that these monomolecular films can only be studied by using the surface balance. There exists no other method through which any direct information can be obtained about molecular packing or interactions.

Monolayer films were studied by using a Teflon trough with a barrier (also of Teflon) that could move across the surface (Figure 4.1). The change in γ was monitored using a Wilhelmy plate attached to a sensor. The accuracy could be as high as mmN/m (m dyne/cm).

The film balance (also called the Langmuir trough) consists of a Teflon (or Teflon-lined) rectangular trough (typically 20 cm × 10 cm × 1 cm). Teflon (PTFE; poly(tetraflouroethylene)) permits keeping the apparatus clean. Clean water (distilled water and purified with active charcoal to remove any minute organic contaminants) is used to fill the trough just over the edge. A barrier of Teflon is placed on one end of the trough, which is used to compress the lipid molecules. Recently, many commercially available film balances are found (monomolecular films).

Lipid (or protein or other film-forming substance) is applied from its solution to the surface. A solution with a concentration of 1 mg/mL is generally used. Lipids are dissolved in $CHCl_3$ or ethanol or hexane (as found suitable). If the surface area of the trough is $100 \, cm^2$, 1–100 μL of this solution may be used. After the solvent has evaporated (about 15 min), the barrier is made to compress the lipid film at a rate of 1 cm/s or as suitable. The amount of lipid or protein applied is generally calculated so as to give a compressed film. Most substances cover $1 \, mg/m^2$ to give a solid film. If there is $100 \, cm^2$ solution in the trough, then 10 μg or less of the film-forming substance is enough for such an experiment. It is thus obvious that such studies can provide much useful surface chemical information with very minute amounts of material. Most proteins can also be studied as monomolecular films (1 mg of protein spreads to cover about $1 \, m^2$ surface area; Birdi, 1999). This method is useful when studying systems where the magnitude of surface pressure (the change in the surface tension) is less than 1 mN/m, and using very high-sensitivity apparatus (± 0.001 mN/m).

Constant area method: One can also study Π versus surface concentration (Cs= area/molecule) isotherms by keeping the area constant. The surface concentration, Cs, is changed by adding small amounts of a substance to the surface (by using a microliter syringe—1 or 5 μL). In general, one obtains a good correlation with the Π versus area isotherms (Gaines, 1966; Adamson and Gast, 1997; Birdi, 1989, 1999).

4.2   MONOLAYER ISOTHERMS ON WATER SURFACES

Lipids with a suitable hydrophilic–lipophilic balance (HLB) are known to spread on the surface of water to form monolayer films. It is obvious that if the lipid-like molecule is highly soluble in water, it will disappear into the bulk phase (as observed for SDS). Thus, the criteria for a monolayer formation are that it exhibits very low solubility in water. The alkyl part of the lipid points away from the water surface. The polar group is attracted to the water molecules and is inside this phase at the surface. This means that the solid crystal, when placed on the surface of water, is in equilibrium with the film spread on the surface. A detailed analysis of this equilibrium has been given in the literature (Gaines, 1966; Adamson and Gast, 1997; Birdi, 2009). The thermodynamics allows one to obtain extensive physical data on this system. It is thus apparent that, by studying only one monolayer of the substance, the effect of temperature can be very evident.

4.3   SELF-ASSEMBLY MONOLAYER (SAM) FORMATION

The most fascinating characteristic some amphiphile molecules exhibit is that, when mixed with water, they form *self-assembly structures*. This was already discussed in Chapter 2 on micelle formation. Since most of the biological lipids also exhibit self-assembly structure formation, this subject has been given much attention in the literature (Birdi, 1999). Lipid monolayer studies thus provide a very useful method to obtain information about SAM formation, both concerning technical systems and cell bilayer structures.

4.3.1   States of Monolayers Spread on Water Surface

It is found that, even a monolayer of lipid (on water), when compressed can undergo various states. In the following text, the various states of monomolecular films will be described as measured from the surface pressure, Π, versus area, $\mathbf{A}$, isotherms, in the case of simple amphiphile molecules. On the other hand, the Π–$\mathbf{A}$ isotherms of biopolymers will be described separately since these have a different nature.

However, before presenting this analysis, it is necessary to consider some parameters of the two-dimensional states, which should be of interest in this context. We need to start by considering the physical forces acting between the alkyl–alkyl groups (parts) of the amphiphiles, as well as the interactions between the polar head groups. In the process where two such amphiphile molecules are brought closer during the Π–$\mathbf{A}$ measurement, the interaction forces would undergo certain changes that would be related to the packing of the molecules in the two-dimensional plane at the interface in contact with water (subphase).

This change in packing is thus analogous conceptually to the three-dimensional P–V isotherms, as is well known in classical physical chemistry (Gaines, 1966; Adamson and Gast, 1997; Birdi, 1989). We know that, as the pressure, P, is increased on a gas in a container, when $T < T_{cr}$, the molecules approach closer, and transition to a liquid phase takes place. Further compression of the *liquid state* results in the formation of a *solid phase*.

In the case of alkanes, the distance between the molecules in the solid phase is ca. 5 Å, while it is 5–6 Å in the case of the liquid phase. The distance between molecules in the gas phase, in general, is ca. $1000^{1/3} = 10$ times larger than in the liquid phase (water: volume of 1 mol water = 18 cc; volume of 1 mol gas = 22.4 L). In fact, monomolecular film studies are the only direct method of obtaining such information at the interfaces of lipids. Considering that only microgram quantities are enough for such information, the importance of such studies becomes clearly evident.

The isotherms of two-dimensional films are also found to resemble the three-dimensional P–V isotherms, and one can use the same classical molecular description as for the qualitative analyses of the various states. However, it is also obvious that it shall not be a complete comparison between the two-dimensional and three-dimensional structures since there are very subtle differences in these two systems, as described in later text.

In the three-dimensional structural buildup, the molecules are in contact with near neighbors as well as with molecules that may be 5 to 10 molecular dimensions apart (as found from x-ray diffraction). This is apparent because in liquids there is a long-range order up to 5–10 molecular dimensions.

On the other hand, in two-dimensional films, the state is much different. The amphiphile molecules are oriented at the interface such that the polar groups are pointed toward water (subphase), while the alkyl groups are oriented away from it. This orientation gives the minimum surface energy. The structure is stabilized through lateral interaction between

1. Alkyl–alkyl group ... *attraction*
2. Polar group–subphase ... *attraction*
3. Polar group–polar group ... *repulsion*

The alkyl–alkyl group attraction arises from the van der Waals forces. The magnitude of van der Waals forces increases with

Increase in alkyl chain length
Decrease in distance between molecules (or when area/molecule [A] decreases)

It means that, as the alkyl chain length increases, the films Π should become more stable, thus giving high collapse pressure Π_{co}. This is also found from experimental data.

Stable films are thus formed when the attraction forces are stronger than the repulsion forces.

4.3.1.1 Self-Assembly Monolayers (SAMs)

Many substances as found in nature (lipids) exhibit unique properties in aqueous media. Some lipids (such as lecithins or alike), when dispersed in water, form very well-defined assemblies, in which the alkyl part of the molecule is in close proximity to each other. This leads to *self-assembly* formation with many important consequences.

Micelle formation is one of the most common SAM structures. In fact, the whole basis of biological cell structure and function is dependent on the lipid-bilayer

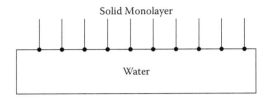

FIGURE 4.3 Orientation of solid films of aliphatic alcohols or acids as films.

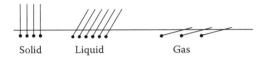

FIGURE 4.4 Lipid monolayer phases.

structure, which is the SAM phenomenon. It has been suggested that, during evolution, life was first conceived by a SAM process.

The most convincing results were those obtained with normal fatty alcohols and acids. Their monomolecular films are stable and exhibit very high surface pressures (Birdi, 1989, 1999). A steep rise in Π is observed around 20.5 Å^2 regardless of the number of carbon atoms in the chains. The volume of a [-CH$_2$-] group is 29.4 Å^3, which gives the length of each [-CH$_2$-] group perpendicular to the surface, or the vertical height of each group, as ca. 1.42 Å. It compares very satisfactorily with the x-ray data, which has a value of 1.5 Å. This means that such straight-chain lipids are oriented in this compressed as vertical (Figure 4.3).

As high pressures lead to transition from gas to liquid or to solid phases in the three-dimensional system, a similar state of affairs would be expected in the two-dimensional film compression Π versus A isotherms (Figure 4.4), as described in the following text.

4.3.1.1.1 Gaseous Films

The most simple type of amphiphile film or a polymer film would be a "gaseous" state. This film would consist of molecules that are at a sufficient distance apart from each other such that lateral adhesion (van der Waals forces) are negligible. However, there is sufficient interaction between the polar group and the subphase that the film-forming molecules cannot be easily lost into the gas phase, and the amphiphiles are almost insoluble in water (subphase).

When the area available for each molecule is many times larger than molecular dimension, the gaseous-type film [state 1] would be present. As the area available per molecule is reduced, the other states, for example, liquid-expanded [L_{ex}], liquid-condensed [L_{co}], and, finally, the solid-like [S or solid-condensed] states would be present.

The molecules will have an average kinetic energy, that is, $1/2\ k_B\ T$, for each degree of freedom, where k is the Boltzmann constant ($=1.372\ 10^{-16}$ erg/T), and T is the temperature. The surface pressure measured would thus be equal to the collisions

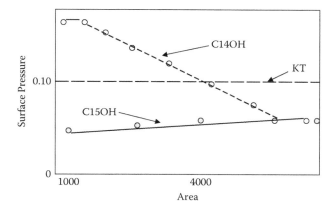

FIGURE 4.5 Π versus **A** plot for an ideal monolayer film: (a) ideal film (Π **A** = 411 = k_B T); (b) $C_{15}OH$; (c) $C_{14}OH$ (Π: mN/m; **A**: Å²/molecule).

between the amphiphiles and the float from the two degrees of freedom of the translational kinetic energy in the two dimensions. It is thus seen that the ideal gas film obeys the relation

$$\Pi\, A = k_B\, T \;[\text{``ideal film''}] \tag{4.1}$$

$$\Pi\,(\text{mN/m})\, A(\text{Å}^2 \text{ per molecule}) = 411\,(T = 298\text{ K})\;[\text{``ideal film''}] \tag{4.2}$$

In Figure 4.5, the Π A versus A is plotted for an ideal monolayer film. Various molecules have been found to give such ideal films (such as $C_{14}H_{29}OH$, valinomycin, proteins). This is analogous to the three-dimensional gas law (i.e., **PV** = kT). At 25°C, the magnitude of (k_B **T**) = 411 10^{16} erg. If Π is in mN/m, and **A** in Å², then the magnitude of k_B **T** = 411. In other words, if one has a system with A = 400 Å²/molecule, then the value of Π = 1 mN/m for the ideal gas film.

In general, *ideal gas* behavior is only observed when the distances between the amphiphiles are very large, and thus the value of Π is very small, that is, <0.1 mN/m. It is also noticed that, from such sensitive data, one can estimate the molecular weight of the molecule in the monolayer. This has been extensively reported for protein monolayers (Adamson and Gast, 1997; Birdi, 1989, 1999).

The latter observation requires an instrument with very high sensitivity, ±0.001 mN/m. The Π versus A isotherms of *n*-tetradecanol (Figure 4.5), pentadecanol, pentadecyclic acid, and palmitic acid are typical data in the low Π region. Similar data for isotherms were reported from other lipid monolayers by other workers.

The various forces that are known to stabilize the monolayers are mentioned as

$$\Pi = \Pi_{kin} + \Pi_{vdw} + \Pi_{electro} \tag{4.3}$$

where
Π_{kin} arises from kinetic forces,

Π_{vdw} is related to the van der Waals forces acting between the alkyl chains (or groups), and

$\Pi_{electro}$ is related to polar group interactions (polar group–water interaction, polar group–polar group repulsion, charge–charge repulsion).

When the magnitude of A is very large, the distance between molecules is large. If there are no van der Waals or electrostatic interactions, then the film obeys the ideal equation. As the area per molecule is decreased, the other interactions become significant. The Π versus **A** isotherm can be used to estimate these different interaction forces. These analytical procedures have been extensively described in the current literature (Gaines, 1966; Adamson and Gast, 1997; Birdi, 1989, 1999).

The ideal equation has been modified to fit Π versus **A** data in those films where the co-area, A_o, correction is needed (Birdi, 1989):

$$\Pi (A - A_o) = k_B T \qquad (4.4)$$

In the case of straight-chain alcohols or fatty acids, A_o is almost 20 Å², which is the same as found from the x-ray diffraction data of the packing area per molecule of alkanes. This equation is thus valid when $A \gg A_o$. The magnitude of Π is 0.2 mN/m for $A = 2000$ Å², for ideal film. However, Π will be about 0.2 mN/m for $A = 20$ Å² for a solid-like film of a straight-chain alcohol.

Π versus **A** for a monolayer of valinomycin (a dodecacyclic peptide) shows that the relation as given in Equation 4.1 is valid (Figure 4.6). In this equation, it is assumed that the amphiphiles are present as monomers. However, if any association takes place, then the measured values of (Π **A**) would be less than $k_B T < 411$, as has also been found (Birdi, 1989, 1999). The magnitude of $k_B T = 4 \cdot 10^{-21}$ J, at 25°C.

In the case of nonideal films, one will find that the versus data does not fit the relation in the equation. This deviation requires that other modified equations-of-state

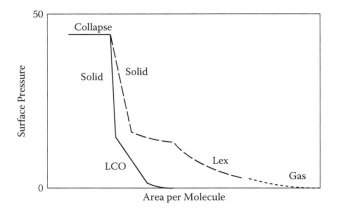

FIGURE 4.6 Π versus $_A$ isotherms for different types of states: (a) gas film; (b) liquid-expanded (L_{ex}) and liquid-condensed (L_{co}); solid films; collapse state.

be used. This procedure is the same as the one used in the case of three-dimensional gas systems.

4.3.1.1.2 Liquid Expanded and Condensed Films

The Π versus A data are found to provide much detailed information about the state of monolayers at the liquid surface. In Figure 4.6, some typical states are shown. The different states have been very extensively analyzed, and will therefore be described in the following text.

In the case of simple amphiphiles (fatty acids, fatty alcohols, lecithins, etc.), in several cases, transition phenomena have been observed between the gaseous and coherent states of films, which show a very striking resemblance to the condensation of vapors to liquids in three-dimensional systems. The liquid films shows various states in the case of some amphiphiles, as shown in Figure 4.6 (schematic). In fact, if the Π versus A data deviate from the ideal equation, then the following interactions may be expected in the film:

Strong van der Waals forces
Charge–charge repulsions
Strong hydrogen bonding with subphase water

This means that the data deviations allow estimation of these interactions.

4.3.1.1.2.1 Liquid Expanded Films (L_{exp}) In general, there are two distinguishable types of liquid films. The first state is called the liquid expanded (L_{exp}) (Gaines, 1966; Chattoraj and Birdi, 1984; Adamson and Gast, 1997). If the Π-A isotherm is extrapolated to zero Π, the value of A obtained is much larger than that obtained for close-packed films, shows that the distance between the molecules is much larger than that in the solid film (to be discussed in later text). These films exhibit very characteristic elasticity.

4.3.1.1.2.2 Liquid Condensed Films (L_{co}) As the area per molecule (or the distance between molecules) is further decreased, a transition to a so-called liquid condensed (L_{co}) state is observed. These states have also been called "solid expanded" films (Adam, 1941; Gaines, 1966; Birdi, 1989, 1999; Adamson and Gast, 1997). The Π versus A isotherms of *n*-pentadecylic acid (amphiphile with a single alkyl chain) have been studied, as a function of temperature (Birdi, 1989). Π-A isotherms for two-chain alkyl groups, such as lecithins, also showed similar behavior.

4.3.1.1.3 Solid Films

As the film is compressed, a transition to a solid film is observed, which collapses at higher surface pressure. The Π versus A isotherms, below the transition temperatures, show the liquid to solid phase transition. These solid films have been also called *condensed films*. They are observed in such systems where the molecules adhere to each other through van der Waals forces very strongly. The Π-A isotherm shows generally no change in Π at high A, while at a rather low A value, a sudden increase in Π is observed, as shown in Figure 4.5. In the case of straight-chain molecules, such as stearyl alcohol, the sudden increase in Π is found to take place at $A = 20 - 22 \text{ Å}^2$, at room temperature (that is, much lower than the phase transition temperature).

From these descriptions, it is seen that the films may, under given experimental conditions, show three first-order transition states, such as (i) transition from the gaseous film to the liquid-expanded (L_{ex}), (ii) transition from the *liquid-expanded* (L_{ex}) to the *liquid-condensed* (L_{co}), and (iii) from L_{ex} or L_{co} to the solid state if the temperature is below the transition temperature. The temperature above which no expanded state is observed has been found to be related to the melting point of the lipid monolayer.

4.3.1.1.4   Collapse States

The measurements of Π versus **A** isotherms generally exhibit, when compressed, a sharp break in the isotherms that has been connected to the *collapse* of the monolayer under given experimental conditions. The monolayer of some lipids, such as cholesterol, is found to exhibit an unusual isotherm (Figure 4.7). The magnitude of Π increases very little as compression takes place. In fact, the collapse state or point is the most useful molecular information from such studies. It has been found that this is the only method that can provide information about the structure and orientation of amphiphile molecules at the surface of water (Birdi, 1989).

However, a steep rise in Π is observed and a distinct break in the isotherm is found at the collapse. This occurs at $Π = 40 \text{ mN/m}$ and $A = 40 \text{ Å}^2$.

This value of A_{co} corresponds to the cholesterol molecule oriented with the hydroxyl group pointing toward the water phase. Atomic force microscope (AFM) studies of cholesterol in Langmuir–Blodgett (LB) films has shown that there exist domain structures (see Chapter X). This has been found for different collapse lipid monolayers (Birdi, 2003). Different data have provided much information about the orientation of lipid on water (Table 4.1).

It is worth mentioning that monolayer studies are the only procedure that allows estimation of the area per molecule of any molecule as situated at the water surface.

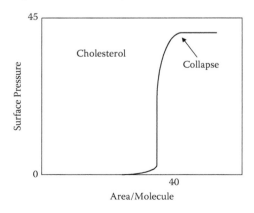

FIGURE 4.7   Π versus **A** isotherm of cholesterol monolayer on water.

TABLE 4.1
Magnitudes of Ao for Different Film-Forming
Molecules on the Surface of Water

Compound	Ao ($\mathring{A}^2$)
Straight-chain acid	20.5
Straight-chain acid (on dilute HCl)	25.1
N-fatty alcohols	21.6
Cholesterol	40
Lecithins	ca. 50
Proteins	ca. 1 m^2/mg
Diverse synthetic polymers (poly-amino acids, etc.)	ca. 1 m^2/mg

Source: Birdi, K. S., *Lipid and Biopolymer Monolayers at Liquid Interfaces*, Plenum Press, New York, 1989; Gaines, 1966; Adamson, A. W. and Gast, A. P., *Physical Chemistry of Surfaces*, 6th ed., Wiley-Interscience, New York, 1997.

In general, the collapse pressure, Π_{col}, is the highest surface pressure to which a monolayer can be compressed without a detectable movement of the molecules in the films to form a new phase (see Figure 4.7).

In other words, this pressure will be related to the nature of the substance and the interaction between the subphase and the polar part of the lipid or the polymer molecule. However, some misinterpretation of the collapse phenomena exists, as found in the older literature.

As described here, the monolayer of a lipid can be formed by different spreading methods. The thermodynamics of the Π_{col} analysis is given in the literature (Birdi, 1989). The monolayer collapse has been shown to provide much information also in the case of protein monolayers.

4.3.1.2   Reaction in Monolayers at Liquid Surfaces

The lipid monolayer is a well-defined structure with the polar part pointing toward the water phase, and the apolar (hydrophobic) part pointing away from the water phase. It is observed that monolayers are very sensitive model to investigate interactions between lipids and other substances (such as ions in the water subphase). The long-chain fatty acids exhibit gas–L_{ex}–solid film states at high pH. However, the same films, on substrates with Ca or Mg ions, exhibit only the solid state. Interestingly, this method may also be used to investigate the kinetics of penetration by measuring the change in Π with time.

ELLIPSOMETRY: The structure of liquid surfaces with monomolecular films can be studied by measuring the light reflected from the surface. The range of thickness that one generally considers to be measured varies from 100 to 1000 $\mathring{A}$ (10–100 nm). However, in monolayers in which the molecules are oriented and the thickness involved is 5–50 $\mathring{A}$, the methods have been not easily pursued. In a differential method in which two beams of light from the same incandescent lamp were directed to two similar troughs, reflected light from each was detected by separate photocells; the photocells are arranged in a sensitive bridge circuit so that very small differences could be determined. With nonabsorbing fatty acid monolayers, the observed change in reflectance were 0.3–3%. Various other reflectance methods have been described by other investigators. The ellipsometers currently in general use have much better sensitivity than earlier models.

4.3.1.3   Fluorescence Studies of Monolayer Films

The technique of fluorescence spectral measurements has become very sensitive over the past decade. In order to obtain more information on the surface monolayers, a new method based on fluorescence was developed. It consisted of placing the monolayer trough on the stage of an epifluorescence microscope, with doped low concentration of fluorescent lipid probe. Later, ordered solid–liquid coexistence at the water–air interface and on solid substrates were reported. The theory of *domain shapes* has been extensively described by this method.

The time-resolved fluorescence of monolayers have been studied by an apparatus described in a recent study (Birdi, 1989). A polarized pulsed beam (20 ps, wavelength 398 nm) from a laser is impinged upon the water surface at an angle of 4 degrees at the surface plane. Fluorescence emission was measured along the surface normal. In these studies, polymers (poly(ethyl)acrylate, poly(ethyl) methacrylate) labeled with anthracene-based fluorescent probe were investigated. The surface pressure versus area isotherms showed collapse at ca. 20 mN/m. The differences in isotherms was correlated to fluorescence data. This method was found to have a sufficiently high sensitivity for studying the mobility of the monolayers at air–water interface. The fluorescence lifetimes were found to fit the presence of two different lifetimes.

4.3.2   OTHER CHANGES AT WATER SURFACES DUE TO MONOLAYERS

The presence of lipid (or similar substance) monolayer at the surface of the aqueous phase gives rise to many changes in the properties at the interface. The major effects that have been investigated extensively are

Surface viscosity (η_s)
Surface potential (ΔV)

There are also other surface properties that change, and these will be described later (Chapter 4.4).

4.3.2.1 Surface Viscosity (ηs)

A monomolecular film is resistant to shear stress in the plane of the surface, as is also the case in the bulk phase: a liquid is retarded in its flow by viscous forces. The viscosity of the monolayer may indeed be measured in two dimensions by flow through a canal on a surface or by its drag on a ring in the surface, corresponding to the "Ostwald" and "Couette" instruments for the study of bulk viscosities. The surface viscosity, η_s, is defined by the relation

$$\{\text{Tangential force per centimeter of surface}\} = \eta_s \times (\text{rate of strain}) \qquad (4.5)$$

and is thus expressed in units of (m t^{-1})(surface poises), whereas bulk viscosity, η, is in units of poise (m^{-1} l t^{-1}). The relationship between these two kinds of viscosity is

$$\eta_s = \eta/d \qquad (4.6)$$

where d is the thickness of the "surface phase," approximately 10^{-7} cm (= 10 Å = 10^{-9} m = nm) for many films. That the magnitude of η_s is of the order of 0.001 − 1 surface poise implies that, over the thickness of the monolayer, the surface viscosity is about 10^4–10^7 poises. This has been compared to the viscosity of butter. The η_s is given in surface poise (gram per second or kilogram per second).

It is easily realized that, if a monolayer is moving along the surface under the influence of a gradient of surface pressure, it will carry some of the underlying water with it. In other words, there is no slippage between the monolayer molecules and the adjacent water molecules. The thickness of such regions has been reported to be of the order of 0.003 cm. It has also been asserted that the thickness would be expected to increase as the magnitude of η_s increases. However, analogous to the bulk phase, the concept of free volume of fluids should be also considered in these films.

Monolayers of long-chain alcohols exhibit η_s approximately 20 times larger than those of the corresponding fatty acids. This difference may explain some data reported on the effect of temperature on monolayers of these molecules (Birdi, 1989). If the monolayer is flowing along the surface under the influence of π, it carries with it some underlying water. This transport is a consequence of the lack of slippage between the monolayer and the bulk liquid adjacent to it. For a monolayer of oleic acid moving at between 1 and 5 cm/s, the direct measurement gives the thickness of the entrained water as approximately 0.003 cm. If the bulk viscosity increases, the thickness of the aqueous layer also increases in direct proportion. Conversely, if the liquid is moving at C, it carries the surface film molecules along with it, giving rise to compression at E, and the backspreading pressure tends to move from E to D. The interfacial water molecules are thus of importance and give rise to these surface phenomena. Accordingly, as mentioned elsewhere, the role of interfacial water needs to be considered in all surface phenomena.

It is also obvious that many such films will exhibit complex viscoelastic behavior, the same as found in bulk phases. The flow behavior then can be treated in terms of viscous and elastic components.

Further, the equilibrium elasticity of a monolayer film is related to the compressibility of the monolayer (analogous to bulk compressibility) by

$$C^s = -1/\mathbf{A} \; (d \; \mathbf{A}/d \; \Pi) \qquad (4.7)$$

where $\mathbf{A}$ is the area per molecule of the film. The surface compressional modulus, K_s (= $1/C^s$), is the reciprocal of C^s.

Since there is no change in surface tension with a change in the rate of a pure liquid surface (i.e., d $\mathbf{A}$/d Π = infinity), the elasticity is zero. The interfacial dilational viscosity, k_s, is defined as

$$\Delta\gamma = k_s \, (1/\mathbf{A}) \, (d \, \mathbf{A}/dt) \tag{4.8}$$

where k_s is the fractional change in area per unit time per unit surface tension difference. From this relation, C^s and E_s can be derived.

Monolayers are thus very useful in understanding various aspects of molecular packing (such as liquid crystals, etc.). With the information from area/molecule, the packing and other interaction parameters can be estimated. These monolayer studies have been found to be important in understanding the thin-liquid film (TLF) structures (bubbles, foams).

4.3.2.2 Surface Potential (ΔV) of Lipid Monolayers

Any liquid surface, especially aqueous solutions, will exhibit asymmetric dipole or ions distribution at the surface as compared to the bulk phase. If SDS is present in the bulk solution, then we will expect that the surface will be covered with SD ions. This would impart a negative surface charge (as is also found from experiments). It is thus seen that the addition of SDS to water not only changes (reduces) surface tension but also imparts negative surface potential. Of course, the surface molecules of methane (in liquid state) will obviously exhibit symmetry in comparison to water molecules. This characteristic can also be associated to the force field resulting from induced dipoles of the adsorbed molecules or spread lipid films (Adamson and Gast, 1997; Birdi, 1989).

The surface potential arises because the lipid molecule orients with polar part toward the aqueous phase. This effects a change in dipole at the surface. There would thus be a change in surface potential when a monolayer is present, as compared to a clean surface. The surface potential, ΔV, is

$$\Delta V = \text{potential}_{monolayer} - \text{potential}_{clean} = V_{monolayer} - V_{clean} \tag{4.9}$$

The magnitude of ΔV is measured most conveniently by placing an air electrode (a radiation emitter Po^{210} [alpha-emitter]), near the surface (ca. millimeter in air), connected to an impedance electrometer. This is required since resistance in air is very high, but it is appreciably reduced by the radiation electrode. However, proper isolation is essential for the data to be reliable.

Since these monolayers are found to be very useful biological cell-membrane structures, it is seen that such studies can provide information on many systems where ions are carried actively through cell membranes (Chattoraj and Birdi, 1984; Birdi, 1999).

The transport of K^+ ions through cell membranes by antibiotics (valinomycin) has been a very important example. Addition of K^+ ions to the subphase of a valinomycin monolayer showed that the surface potential became positive. This clearly indicates the ion-specific binding of K^+ to valinomycin (Birdi, 1989).

4.3.3 CHARGED LIPID MONOLAYERS

The spread monolayers have provided much useful information about the role of charges at interfaces. In the case of an aqueous solution consisting of fatty acid or

SDS, R-Na, and NaCl, for example, the Gibbs equation [2.40] may be written as (Adamson and Gast, 1997; Birdi, 1989; Chattoraj and Birdi, 1984):

$$-d\gamma = \Gamma_{RNa}\, d\mu_{RNa} + \Gamma_{NaCl}\, d\mu_{NaCl} \tag{4.10}$$

Further,

$$\mu_{RNa} = \mu_R + \mu_{Na} \tag{4.11}$$

$$\mu_{NaCl} = \mu_{Na} + \mu_{Cl} \tag{4.12}$$

It can be easily seen that the following will be valid:

$$\Gamma_{NaCl} = \Gamma_{Cl} \tag{4.13}$$

and

$$\Gamma_{RNa} = \Gamma_R \tag{4.14}$$

It is also seen that following equation will be valid for this system:

$$-d\gamma = \Gamma_{RNa}\, d\mu_{RNa} + \Gamma\, d\mu_{Na} + \Gamma\, d\mu_{Cl} \tag{4.15}$$

This is the form of the Gibbs equation for an aqueous solution containing three different ionic species (e.g., R, Na, Cl). Thus, the more general form for solutions containing i number of ionic species would be

$$-d\gamma = \Sigma\, \Gamma_i\, d\mu_i \tag{4.16}$$

In the case of charged films, the interface will acquire *surface charge*. The surface charge may be positive or negative, depending on the cationic or anionic nature of the lipid or polymer ions. This would lead to the corresponding *surface potential*, ψ, also having a positive or negative charge (Chattoraj and Birdi, 1984; Birdi, 1989). The interfacial phase must be electroneutral, which can only be possible if the inorganic counterions also are preferentially adsorbed in the interfacial phase (Figure 4.8).

The surface phase can be described by the Helmholtz double-layer theory (Figure 4.8.) If a negatively charged lipid molecule, R-Na$^+$, is adsorbed at the interface AA′, the interface will be negatively charged (air–water or oil–water). According to the Helmholtz model for the double layer, Na$^+$ on the interfacial phase will be arranged in a plane BB′ toward the aqueous phase. The distance between the two planes, AA′ and BB′, is given by Ù. The charge densities are equal in magnitude but with opposite signs, Γ (charge per unit surface area), in the two planes. The (negative) charge density of the plane AA′ is related to the surface potential (negative), ψ_o, at the Helmholtz charged plane:

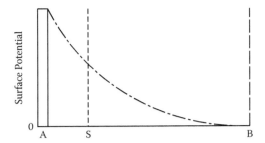

FIGURE 4.8 Double-layer model (schematic). (See text for details.)

$$\psi_o = [4 \pi \sigma \delta]/\mathbf{D} \tag{4.17}$$

where $\mathbf{D}$ is the dielectric constant of the medium (aqueous). According to the Helmholtz double-layer model, the potential ψ decreases sharply from its maximum value, ψ_o, to zero as δ becomes zero (Birdi, 1989, 1999). The variation of ψ is depicted in Figure 4.9.

The Helmholtz model was found not to be able to give a satisfactory analysis of measured data. Later, another theory of the diffuse double layer was proposed by Gouy and Chapman. The interfacial region for a system with charged lipid, R-Na$^+$, with NaCl, is shown in Figure 4.10.

As in the case of the Helmholtz model, the plane AA' will be negative due to the adsorbed R-species. Therefore, the Na$^+$ and Cl$^-$ ions will be distributed nonuniformly due to electrostatic forces. The concentrations of the ions near the surface can be given by the Boltzmann distribution, at distance x from the plane AA', as

$$C^s_{Na+} = C_{Na+} [\exp - (\epsilon\psi/kT)] \tag{4.18}$$

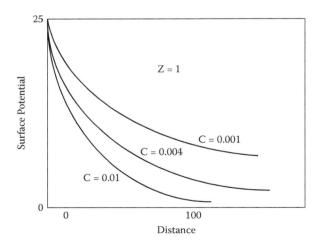

FIGURE 4.9 Double-layer model (variation of ψ with distance).

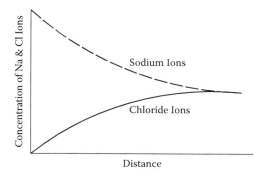

FIGURE 4.10   Variation of Na$^+$ and Cl$^-$ ions near the charged surface.

$$C^s_{Cl-} = C_{Cl-} [\exp + (\varepsilon\psi/kT)] \tag{4.19}$$

where C_{Na+} and C_{Cl-} are the number of sodium and chloride ions per milliliter, respectively, in the bulk phase. The magnitude of ψ varies with x as shown in Figure 4.10, from its maximum value, ψ_o, at plane AA′. From the foregoing equations, we find that the quantities C^s_{Na+} and C^s_{Cl-} will decrease and increase, respectively, as the distance x increases from the interface, until their values become equal to C_{Na+} and C_{Cl-}, where ψ is zero. The variation of C^s_{Na+} and C^s_{Cl-} with x are given in Figure 4.10. The extended region of x between AA′ and BB′ in Figure 4.10 may be termed the diffuse or the Gouy–Chapman double layer.

The volume density of charge [per milliliter] at a position within AA′ BB′ may be defined as equal to

$$\varepsilon [c^{s+} - c^{s-}] \tag{4.20}$$

can be expressed by the Poisson relation

$$d^2 \psi/d^2 x = -[4 \pi]/D \tag{4.21}$$

In this derivation, it is assumed that the interface is flat, such that it is sufficient to consider only changes in ψ in the x direction normal to the surface plane.

The following relation can be derived from this:

$$[d5 \psi/d x5] = d[d\psi/dx]/dx \tag{4.22}$$
$$= -[4 \pi N \varepsilon/1000 D] c$$

$$[e^{-\varepsilon/kT} e^{+\varepsilon\psi/kT}] \tag{4.23}$$

where c is the bulk concentration of the electrolyte.

In a circle of unit surface area on the charged plane A-A′, the negative charges acquired by the adsorbed organic ions (amphiphiles) within this area represent the surface charge density σ:

$$\sigma = -\int \rho \, d \, x \tag{4.24}$$

$$= [D/4 \, \pi][d \, \psi/d \, x] \tag{4.25}$$

when integration is zero and infinity. At $x = 0$, the magnitude of ψ reaches ψ_o:

$$\sigma = [2 \, D \, R \, T \, c/1000 \, \pi]^{1/2} \, [\sinh (\varepsilon \, \psi o/2 \, k \, T)] \tag{4.26}$$

The average thickness of the double layer, $1/k$, that is, the Debye–Huckel length, is given as (Chattoraj and Birdi, 1984):

$$1/k = [1000 \, D \, R \, T/8 \, \pi \, N5 \, \varepsilon 5 \, c]^{0.5} \tag{4.27}$$

At 25°C, for uni-univalent electrolytes, one gets

$$k = 3.282 \ 10^{\,7} \, c^{-1} \, [\text{cm}] \tag{4.28}$$

For small values of ψ, one gets the following relationships:

$$\sigma = [\mathbf{D} \ \mathbf{R} \ \mathbf{T} \, k/2 \, \pi \, N \, \varepsilon] \, \sinh[\varepsilon \, \psi_o/2 \, k \, \mathbf{T}] \tag{4.29}$$

This relates the potential charge of a plane plate condenser to the thickness $1/k$. The expression based upon the Gouy model is derived as

$$\sigma = 0.3514 \ 10^5 \, \sinh \, [0.0194 \, \psi_o] \tag{4.30}$$

$$= \Gamma \, z \, N \, \varepsilon \tag{4.31}$$

where the magnitude of Γ can be experimentally determined, and thus the magnitude of ψ_o can be estimated. The free energy change due to the electrostatic work involved in charging the double layer is (Adamson and Gast, 1997; Chattoraj and Birdi, 1984; Birdi, 1989)

$$Fe = I^{\,\psi o} \, \sigma \, d \, \psi \tag{4.32}$$

By combining these equations, the expression for Π_{el} can be written (Chattoraj and Birdi, 1984):

$$\Pi_{el} = 6.1 \ c^{1/2} \, [\cosh \, \sinh^{-1} \, (134/A_{el} \, C^{1/2})]^{-1} \tag{4.33}$$

The quantity $[k \, T]$ is approximately $4 \ 10^{-14}$ erg at ordinary room temperature (25°C), and $[k \, T/\varepsilon] = 25$ mV. The magnitude of Π_{el} can be estimated from monolayer studies at varying pH. At the isoelectric pH, the magnitude of Π_{el} will be zero (Birdi, 1989). These Π versus A isotherms data at varying pH subphase have been used to estimate Π_{el} in different monolayers.

4.3.3.1 Transport of Ions in Biological Cell Membranes

The most important biological cell-membrane function is the transport of ions (such as Na, K, Li, Mg) through the hydrophobic lipid part of the bilipid membrane (BLM) (Birdi, 1989, 1999). This property is related to many conditions, such as infection, and thus to the activity of antibiotics. These complicated biological processes have been studied using monolayer model systems. For instance, valinomycin monolayers have been extensively investigated. They exhibited K-ion specificity exactly as found in biological cells. It is well known that cell membranes inhibit the free transport of ions (the alkyl chains of the lipids hinder such transport). However, molecules such as valinomycin assist in specific ion (K-ion) transport through binding (Birdi, 1989).

4.4 EFFECT OF LIPID MONOLAYERS ON EVAPORATION RATES OF LIQUIDS

In arid and semiarid areas, the amount of water lost from reservoirs by evaporation frequently exceeds the amount beneficially used. Evaporation of water from lakes and other reservoirs is a very important phenomenon in those parts of the world where water is not readily available. Further, from ecological considerations (rain and temperature), the evaporation phenomenon has much importance for global temperature. Cloud formation depends on the evaporation of water from lakes and oceans. Clouds are thus isolators for sunlight reaching Earth. This cycle is related to the global heating process, and has been extensively discussed with regard to CO_2 and global warming. The reduction of even a part of the evaporation losses would therefore be of incalculable value.

In general, the fall in water level would arise from the following parameters:

Fall in water level + rainfall = evaporation + seepage + abstractions

At the air–water interface, water molecules are constantly evaporating and condensing in a closed container. In an open container, water molecules at the surface will desorb and diffuse into the gas phase. It is therefore important to determine the effect of a monomolecular film of amphiphiles at the interface. The measurement of the evaporation of water through monolayer films was found to be of considerable interest in the study of methods for controlling evaporation from great lakes. Many important atmospheric reactions involve interfacial interactions of gas molecules (oxygen and different pollutants) with aqueous droplets of clouds and fog as well as ocean surfaces. The presence of monolayer films would thus have an appreciable effect on such mass transfer reactions.

In the original procedure for measuring the evaporation of water, the box containing the desiccant is placed over the water surface, and the amount of water sorbed is determined by simply removing the box and weighing it (La Mer, 1962; Langmuir, 1943). The results are generally expressed in terms of specific evaporation resistance, **r**. The methods for calculating **r** from the water uptake values, together with the assumptions involved, are described in detail in the aforementioned references. The rates of evaporation are measured both without (Rw) and with (Rf) the film. The resistance **r** is given as

$$\mathbf{r} = A\ (\mathbf{vw} - \mathbf{vd})\ (Rf - Rw) \qquad (4.34)$$

where A is the area of the dish, and **vw** and **vd** are the water vapor concentrations for water and desiccant, respectively. The condensed monolayers gave a much higher **r** value than the expanded films, as expected.

A more advanced arrangement was described that made use of recording film balances, in which the temperature difference between two cooled sheet metal probes was measured. If both probes are over clean water and the rate of moisture condensation is the same, then there is no difference in the temperature. However, if monolayer is present, the retardation of evaporation gives rise to a temperature difference.

Other techniques for the diffusion of gas through monolayers at the liquid interface have also been investigated (Blank, 1970). In these methods, the differential manometer system was used to measure the adsorption of gases such as CO_2 and O_2 into aqueous solutions with and without the presence of monolayers. The Geiger–Mueller counter with a suitable sorbent and a radioactive tracer gas was used to measure the reduction of evolution of H_2S and CO_2 from the surface solution when a monolayer was present.

Many decades ago, it was found that the evaporation rates of water were reduced by the presence of monomolecular lipid film. This observation resulted in important applications in reducing the loss of water from great lakes during summer in such continents as Australia and Africa. The simplest method is to measure the loss of weight of a water container with or without the presence of a monomolecular lipid film. La Mer investigated the effect of long-chain films ($C_{14} H_{29}$ OH, C_{16} OH, C_{18} OH, C_{20} OH, C_{22}OH), and found that the resistance to evaporation increased with chain length. For instance, the resistance increased by a factor 40 for C_{14}OH to C_{22}OH. This indicates that evaporation takes place through the alkyl chains. Since the attraction between alkyl chains increases with the number of carbon atoms, the resistance to evaporation also increases.

4.4.1 MIXED MONOLAYERS OF LIPIDS

The spread mixed lipid monolayer studies provide information about the packing and orientation of such molecules at the water interface. These interfacial characteristics affect many other systems. For instance, mixed surfactants are used in froth flotation. The monolayer surface pressure of a pure surfactant is measured after the injection of the second surfactant. From the change in Π, the interaction mechanism can be measured. The monolayer method has also been used as a model biological membrane system. In the latter BLM, lipids are found to be mixed with other lipid-like molecules (such as cholesterol). Hence, mixed monolayers of lipids + cholesterol have been found to provide much useful information on BLM. The most important BLM and temperature melting phenomena is the human body temperature regulation. Normal body temperature is 37°C (98°F), at which all BLM function efficiently.

In biological systems, the lung fluid exhibits surface pressure characteristics that are related to its lipid composition. The ratio between two different lipids has been shown to be critical for lung function in newborn babies. Recent studies of mixed monolayers have been made using AFM (Birdi, 2002a).

4.5 MONOLAYERS OF MACROMOLECULES AT THE WATER SURFACE

Many macromolecules (such as proteins) form stable monolayers at the water surface if the hydrophilic–lipohilic balance (HLB) is of the right quantity. Especially, almost all proteins (hemoglobin, ovalbumin, bovine serum albumin, lactoglobulin, etc.) are reported to form stable monolayers at the water surface.

4.6 LANGMUIR–BLODGETT (LB) FILMS (TRANSFER OF LIPID MONOLAYERS ON SOLIDS)

Some decades ago (Langmuir, 1920), it was reported that, when a clean glass plate was dipped into water covered by a monolayer of oleic acid, an area of the monolayer equal to the area of the plate dipped was deposited on withdrawing the plate. Later, it was found that any number of layers could be deposited successively by repeated drippings, later called the Langmuir–Blodgett (LB) method (Petty, 1996; Roberts, 1990; Birdi, 1989).

In another context, the electrical properties of thin films obtained by different procedures, for example, thermal evaporation in vacuum, have been investigated in much detail. However, the films deposited by the LB technique have only recently been used in electrical applications. Thickness in LB films can be varied from only one monomolecular layer (ca. $25\text{Å} = 25 \ 10^{-10}\,\text{m}$), while this is not possible by evaporation procedures.

Monomolecular layers (LB films) of lipids are of interest in a variety of applications, including the preparation of very thin controlled films for interfaces in solid-state electronic devices (Gaines, 1966; Birdi, 1989, 1999). Langmuir (1920) investigated the process of transferring the spread monolayer film to a solid surface by raising the solid surface through the interface. The process of transfer is depicted (schematic) in Figure 4.11. It is seen that, if the monolayer is in a closely packed state, then it is transferred to the solid surface, most likely without any change in the packing density. Detailed investigations have, however, shown that the process of transfer is not as simple as shown in Figure 4.11. The monolayer may or may not be stable on a solid surface, and defects may be present. However, by using modern AFM techniques, one can determine the molecular orientation and packing of such LB films (Birdi, 1989, 1999, 2002, 2002a; Gaines, 1966).

Scientists are currently using LB film assemblies as solutions to problems in diverse areas such as microlithography, solid-state polymerization, light guiding, electron tunneling, and photovoltaic effects. In the case of such films as Mg stearate, if a clean glass slide is dipped through the film, a monolayer is adsorbed on the downstroke. Another layer is adsorbed on the upstroke. Under careful conditions,

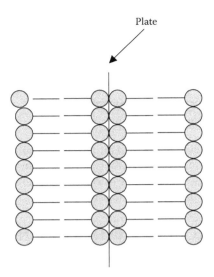

FIGURE 4.11   Transfer of monolayers of lipids to LB films.

LB films with multilayers (varying from a few layers to thousands) could be made, and the adsorption monitored by measuring the decrease in Π on each stroke. If no adsorption takes place, then no change is observed. There are some lipids, such as cholesterol, which do not form LB films. There are also other methods, such as

Light reflection
IR spectroscopy
STM (scanning tunneling microscope) and AFM (atomic force microscope)
Change in contact angle

that can provide a detailed information on these LB films.

The LB deposition is traditionally carried out in the "solid" phase. The surface pressure is then high enough to ensure sufficient cohesion in the monolayer; for example, the attraction between molecules in the monolayer is high enough so that the monolayer does not fall apart during transfer to the solid substrate. This also ensures the buildup of homogeneous multilayers. The surface pressure value that gives the best results depends on the nature of the monolayer and is usually established empirically. However, amphiphiles can seldom be successfully deposited at surface pressures lower than 10 mN/m, and at surface pressures above 40 mN/m, collapse and film rigidity often pose problems. When the solid substrate is hydrophilic (glass, SiO_2, etc.), the first layer is deposited by raising the solid substrate from the subphase through the monolayer, whereas if the solid substrate is hydrophobic (HOPG, silanized SiO_2, etc.), the first layer is deposited by lowering the substrate into the subphase through the monolayer. It is found that, in some LB systems, the magnitude of Π drops about 1–2 mN/m as each time a plate (with surface area of 5 cm^2) is moved down or up through the monolayer. It is thus a very sensitive method to study the LB deposition phenomenon directly.

There are several parameters that affect on what type of LB film is produced. These are the nature of the spread film, subphase composition and temperature, surface pressure during deposition and deposition speed, the type and nature of the solid substrate, and the time the solid substrate is stored in air or in the subphase between deposition cycles.

The quantity and quality of the deposited monolayer on a solid support is measured by a so-called transfer ratio, t_r. This is defined as the ratio between the decrease in monolayer area during a deposition stroke, Al, and the area of the substrate, As. For ideal transfer, the magnitude of t_r is equal to 1. Depending on the behavior of the molecule, the solid substrate can be dipped through the film until the desired thickness of the film is achieved. Different kinds of LB multilayers can be produced and/or obtained by successive deposition of monolayers on the same substrate (see Figure 4.11). The most common one is the Y-type multilayer, which is produced when the monolayer deposits on the solid substrate in both up and down directions. When the monolayer deposits only in the up or down direction, the multilayer structure is called either **Z**-type or **X**-type. Intermediate structures are sometimes observed for some LB multilayers, and they are often referred to be **XY**-type multilayers.

The production of the so-called alternating layers that consist of two different kind of amphiphiles is also possible by using suitable instruments. In such an instrument, there is a trough with two separate compartments, both possessing a floating monolayer of a different amphiphile. These monolayers can then be alternately deposited on one solid substrate.

The monolayer can also be held at a constant surface pressure, which is enabled by a computer-controlled feedback system between the electrobalance and the motor responsible for the movements of the compressing barrier. This is useful when producing LB films, that is, when the monolayer is deposited on a solid substrate.

The most simple procedure generally used is one in which a clean and smooth solid surface (of suitable surface area) is dipped through the interface with the monolayer. Alternatively, one can also place the solid sample in the water before a monolayer is spread and then drawn up through the interface to obtain the film transfer.

It is obvious that such processes involving monomolecular film transfers will easily be disturbed by defects arising from various sources. As will be shown in the following text, these defects are in most cases easily detected. The structural analysis of the molecular ordering within a single LB monolayer is important both to understand how the environment in the immediate vicinity of the surface (i.e., solid) affects the structure of the molecular monolayer and to ascertain how the structure of one layer forms a template for subsequent layers in a multilayer formation.

Studies of the order within surfactant monolayers have been reported for many decades. Multilayer assemblies have been studied by electron as well as infrared absorption. Motivated by an older model proposed for the orientation of molecules (Langmuir, 1933; Epstein, 1950), and by recent theoretical calculations, these two potential models for tilt disorder in the monolayer have been examined. Both models arise because the monolayer structure tries to compensate for the difference between the equilibrium head–head and chain–chain distances that each piece of the molecule would want to attain if it were independent. In one model, the magnitude of the tilt is fixed, but the tilt direction wanders slowly through the lattice. In the second model, the tilt always points to the center of the cluster, but the magnitude varies from straight up at center of the cluster to some maximum tilt at the cluster edge. Such a structure contains display distortion of the tail ordering and has been called "micellar cluster." Both models exhibit contributions to translational disorder beyond that existing in the head group lattice. The micellar cluster model preserves strong bond orientational correlations, while the first model degrades them to some extent. Simulations have shown that the micellar model can fit the general features of the diffraction pattern better than the first model. The precise nature of the positional disorder, liquid-like or microcrystalline, is difficult to determine for the strongly disordered structure of complex molecules found in the LB monolayer. Especially, one would like to know if the structure is similar to a stacked-hexatic phase (long-range bond orientation with exponentially decaying translational correlations) or microcrystalline (perfect translational correlations).

Electrical behavior of LB films: Insulating thin films in the thickness range 100–20000 Å (100–20000 10^{-10} m) have been the subject of varied interest among the scientific community because of the potential applied significance for developing devices, such as optical, magnetic, electronic, etc. Some of the unusual electrical properties possessed by thin LB films that are unlike those of bulk materials led to the pursuit of their technological applications, and, consequently, interest in thin-film studies grew rapidly. Earlier studies did not prove to be very inspiring because the LB films obtained always suffered from the presence of pinholes, stacking faults, and other impurities, and hence, the results were not reproducible. It is only in the past few decades that many sophisticated methods have become available for the production and examination of thin films, and reproducibility of the results could be and controlled to a greater extent. Nevertheless, the unknown nature of inherent defects and a wide variety thin-film systems still complicate the interpretation of many experimental data and thus hinder their use in devices.

Breakdown conduction in thin films, the major subject of investigation, has been based on the films prepared by thermal evaporation under vacuum or similar techniques. It was realized that LB films have remained less known among the investigators of this field. Their various interesting physical properties have been investigated in the current literature.

4.6.1 STRUCTURE AND ORIENTATION OF ORGANIZED MONOLAYER ASSEMBLIES

The structure and orientation of the deposited amphiphile molecules have been found to be governed by the angle of contact between the monolayer and the solid surface. The deposited monolayers, in general, have been characterized as "X-," "Y-," and "Z"-type, and their molecular arrangements can be described as follows.

In the case of X and Z-type films, the molecules are oriented in the same direction, and thus, the surface will be composed of carboxyl and methyl groups. On the other hand, in a Y-type film, the molecules in adjacent layers will be oriented in opposite directions, and the LB film surface will be composed of methyl groups. Of these three types of LB films, the one that has been studied is the Y-type, in which the monolayer transfer takes place both ways on each dipping and withdrawal of the solid slide across the interface. On the other hand, in X-type films, transfer takes place only when the slide is dipped, and in the Z-type, only when the slide is withdrawn. The orientation is termed *exotropic* when the methyl groups touch the solid surface and the polar groups (carboxyl) remain away from it in the first layer. If the orientation of these two groups is reversed, it is termed *endotropic*. Thus, an X-type film is made up of a series of exotropic layers, a Y-type a series of alternating exotropic and endotropic layers, and an Z-type a series of endotropic layers. Z-type films are, rather uncommon.

There are two crucial factors for making satisfactory electrical measurements on the LB films uniform packing and thickness. These two requirements are generally fulfilled to a greater extent when one uses LB films. Metrical thicknesses of several fatty acid monolayers have been measured using the best-known optical methods. A high quality of fringes and, accordingly, high-precision measurements were obtained employing a protective colloid ion layer.

As regards uniformity in the packing of LB films, the selection and purity of the substrate needs to be carried out very carefully. In fact, to achieve the required uniformity in the packing of LB films, the deposition of the first monolayer has been found to be very critical since any voids or imperfections in it generally leads to major disruption of the subsequent monolayers. Therefore, it has been suggested that the solid slide emerging from the liquid surface must be completely dried before being reimmersed. If the wet slide is reimmersed, generally no multilayers are adsorbed (for lecithin, fatty acids, and protein monolayers).

Ba- or Cu-stearate films could be built up to as many as 3000 layers simply by repeated dipping and withdrawal processes (Birdi, 1989, 1999). On the other hand, fatty-acid films or other sources have cracking and fogging tendencies. It was suggested that addition of the Cu ion [10^{-6} M] solved these problems (Birdi, 1999).

4.6.2 OPTICAL AND RELATED METHODS OF ABSORPTION

Optical reflection from a monomolecular film must be measured from the interface with a very small amount of material present. Therefore, in these methods, repeated interaction of the light beam with one or more identical films is generally employed. The simplest way to observe a light beam that has passed through several identical monolayers is to transfer portions of the layer to suitable transfer end plates, which are then stacked and examined in a conventional spectrophotometer. This method was used where monolayers of chlorophyll were deposited on glass slides by the LB method. Ferrodoxin and chlorophyll monolayers were investigated by measuring the spectra (550–750 nm) of these films at the interface.

To examine monolayers on liquid surfaces in situ, multiple interactions by mounting mirrors above and below the surface have been used. X-ray reflectivity measurements have been used to study the counterion overlayers at the interface between electrolyte solutions and monolayers of carboxylic acids terminated alkanethiols self-assemblies on Au (Birdi, 1999).

The Brewster angle microscope (BAM) has been used to investigate the monolayers of polyamic acid at the air–water interface in situ and without the addition of probes. The morphology of the monolayer was observed as a function of surface pressure. Nucleation growth has been studied by fluorescence (probe) and BAM (without probe) methods. In the case of lipid monolayers with probes added, the effect of probes on the monolayer structures may be questioned. Nucleation and growth of three-dimensional crystallites in fatty acid monolayers at constant pressure was investigated by BAM. The growth of domains could be followed by BAM, and various shapes ranging from dots and triangles to thread-like strips were observed.

Grazing incidence x-ray diffraction (GID) measurements have indicated that both precollapse and collapsed state monolayers at the air–water interface can be crystalline (Birdi, 1989). A general procedure was delineated that could provide near-atomic resolution of two-dimensional crystal structures of n-triacontanoic acid ($C_{29}H_{59}COOH$). A monolayer composed of rod-like molecules would generally pack in such a way that each molecule has six nearest neighbors, that is, hexagonal cell.

4.6.3 LB-Deposited Film Structure

No direct method exists by which monolayer film structures on water can be studied. Therefore, the LB method has been used to study molecular structures in past decades. The most useful method for investigating the detailed LB-deposited film structure is the well-known electron diffraction technique (or the scanning probe microscope [Birdi, 2002a]). The molecular arrangements of deposited mono- and multilayer films of fatty acids and their salts, using this technique, have been reported. The analyses showed that the molecules were almost perpendicular to the solid surface in the first monolayer. It was also reported that Ba-stearate molecules have a more precise normal alignment compared to stearic-acid monolayers. In some investigations, the thermal stability of these films has been found to be remarkably stable up to 90°C.

Based on structural analyses obtained by the electron diffraction technique, the deposited films are known to be monocrystalline in nature, and thus, can be regarded as a special case of a layer–bilayer mechanical growth forming almost "two-dimensional" crystals. However, there is evidence that Ba-behenate multilayers do in fact show absence of crystallization, as demonstrated by electron micrographic studies.

In many of the early reports, it has been shown that deposited films obtained by the usual process of monolayer transfer invariably contained holes, cracks, or similar imperfections. These are not surprising because, at higher Π, the more compact film will be removed. Nevertheless, it would be an oversimplification to regard the film transferred at higher Π as perfectly uniform, coherent, and defect free. Artifacts are indeed introduced if proper care has not been taken either during the transfer process, and the subsequent thermal evaporation of metal electrodes over the film may disturb the film structure. In some cases, radio autographic investigations have indicated that these deposited layers are generally uniform with no apparent gross defects. It is thus found that recent modifications in deposition technique have made it possible to obtain films largely free from gross defects or imperfections. However, defects or pinholes are easily detected since resistance is very low around them.

In a recent study, LB films of iron stearate was deposited from aqueous solution into a substrate only on the down journey. On the up journey, the substrate was withdrawn through a clean area. The condition of the aqueous surface was the same as that for the preparation of the Y-type layer. The Fe55 tracer techniques was used for the examination of the direction of molecules. The unidirectional surface conductance of monolayers of stearic acid deposited on a glass support was investigated.

The influence of two-dimensional pressure on the tangential permeability of the layers for various cations was examined. The current in the channel formed by the monomolecular film and the substrate is carried by hydrated cations. The permeability of the channel to various cations diminishes in the order

$$Ca^{++}, Zn^{++}, Ba^{++}, Mg^{++}, H^+, Cs^+, K^+, Na^+, Li^+$$

The contact angles and adhesional energy changes during the transfer of monolayers from the air–water interface to solid (hydrophobic glass) supports have been analyzed (Gaines, 1966; Birdi, 1989).

4.6.3.1 Molecular Orientation in Mixed-Dye Monolayers on Polymer Surfaces

In the past few years, there has been an increasing interest in the use of surface-active dyes to study the properties of biological and artificial membranes and to construct monomolecular systems by self-organization. When these dyes are incorporated in lamellar systems, it is found that the paraffin chains stand perpendicularly on the plane of the layer, while the chromophores lie flat near the hydrophilic interface. In order to develop new molecules as functional components of monolayer assemblies, a series of nine surface-active azo and stilbine compounds were synthesized. Their monolayer properties at the air–water interface were investigated by Π and spectroscopic techniques. The adsorption on stretched polyvinyl alcohol (PVA) films of mixed monolayers of n-octadecyltrichlorosilane [OTS-$C_{18}H_{37}SiC_{13}$] plus long-chain-substituted cyanine dyes were investigated. These systems were selected because of

i. The solid support possesses a simple organization with uniaxial symmetry, which allows a straightforward correlation with the molecular orientation induced in the monolayer.
ii. The solid substrate (PVA) is transparent, and the adsorbed molecules contain elongated dye chromophores, so that the molecular orientation within the monolayer can be readily determined by means of polarized absorption spectroscopy (linear dichroism [LD]).
iii. It is easy to produce the support in large quantities, and adsorption is easy for oleophobic octadecyltricholorosilane (OTS) monolayers.

Furthermore, PVA is not soluble, and does not swell in the organic solvents used in monolayer studies. The orientational effects were estimated from LD.

An application of neutron activation analysis for the determination of inorganic ions in LB multilayers was reported. A special technique for the removal of multilayers from the solid substrate was used. Multilayers of arachidic acid and octa-decylamine were analyzed with respect to Cd^{++} and HPO_4-.

Interest in the dielectric studies of deposited LB films of fatty acids and their metal salts was one of the parameters of main investigation in the early stages of research on LB films, for example, capacitance, resistance, and dielectric constant. In early investigations, measurements on impedance of films and related phenomena were carried on Cu- and Ba-stearate, and Ca-stearate using both X- and Y-type films. Initially, Hg droplets were used for small-area probe measurements and an AC bridge was used for impedance measurements. The resistance of the films was found to be very low (<1 Ω), with high signal voltages, whereas it was of the order of megaohms with signals of 1 or 2 V. The results were not satisfactory, and later measurements were carried out after replacing the AC bridge by a radiofrequency bridge. The measurements at frequencies of 1 MHz and 0.244 MHz and for films containing 7–41 deposited monolayers determined dielectric constant values ranging between 1.9 and 3.5 with an average value of 2.5. In both types of films, the capacity decreased with thickness, as can be expected from the following relation:

Capacitance of the deposited LB films:

$$C = A \, \varepsilon/4 \, \pi \, N \, d \qquad (4.35)$$

where C is the capacity, ε denotes the dielectric constant, N is the number of layers, d the layer thickness, and A the area of contact between drop and film. On the other hand, the values of the resistance per layer showed a definite increase with the thickness of the film. The specific resistance of the films, thus determined from their values of the resistance per layer, was ca. 10^{13} ohms. This was based on the results of capacitance measurements on some 75 samples. The measurements thus performed on stearate films (1–10 layers) led to ε values between 2.1 and 4.2, with a bulk value of 2.5.

In many of the measurements reported in the literature, the organic film was sandwiched between evaporated aluminum electrodes. The fact that an oxide layer grows on the base of the aluminum electrode was present, and its effect on the capacitance values of the device was neglected considering that the resistivity of oxide film is small compared to that of the organic LB layers. The presence of such a thin oxide layer between metal electrodes and fatty acids can be analyzed, and the capacitance has been reported to be a linear function of [1/C] with respect to the number of transferred monolayers (Figure 4.12). LB films of Ba-salts of fatty acids deposited at $\Pi = 50$ mN/m (Birdi, 2002a) gave the following relation between [1/C] and N:

Ba-stearate:

$$1/C = 15.9 \, N + 1.13, \, [10^6 \, F^{-1}] \qquad (4.36)$$

Ba-behenate:

$$1/C = 17.2 \, N + 8, \, [10^6 \, F^{-1}] \qquad (4.37)$$

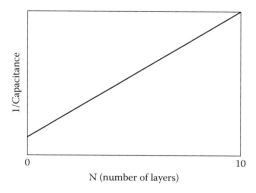

FIGURE 4.12   Variation of [1/C] (C = capacitance) of LB films versus number of layers (**N**) (schematic).

The dielectric anisotropy of long-chain fatty acid monolayers was analyzed. These fatty acids were considered as being oriented in a cylinder cavity with length (L) >> diameter (D). Considering each bond in these molecules as a polarization ellipsoid with axial symmetry about the -C-C- bonds, the mean polarizibtlity of the bonds was calculated.

The LB film of fatty-acid salt has been known for long time to have breakdown strength ca. 10^6 V/cm. The earliest study of breakdown of the films under high DC voltages were measured using a galvanometer to read the current. Using an Hg drop as the upper electrode, no current was detected, until the magnitude of the applied voltage across the specimen reached critical value in the case of a 30-layer LB film, whereas in thinner films, appreciable currents were present almost from start. The breakdown per layer was found to rise sharply at a thickness of ca. 20 layers when its order of magnitude was ca. 10^6 V/cm. The specific resistance was found to be ca. 10^{13} ohms, below the breakdown voltage. In later studies, water drop with electrolyte was used instead of Hg, and no difference was found between the two methods. The destructive breakdown in LB films has been found to be composed of two events: the destructive breakdown voltage and the maximum breakdown voltage. These events have been determined in various fatty-acid soaps in the thickness range corresponding to 16–80 layers.

The destructive breakdown voltage has been reported to be independent of the material and thickness of the film, and the maximum breakdown voltage has been reported to vary slightly with increase film thickness. The breakdown voltage system has been reviewed by various investigators. The breakdown mechanism in LB film has been described in terms of a statistical model (Agarwal, 1974; Birdi, 1999). The AC breakdown in LB film is still at a stage where it is not well understood. The study of AC voltage characteristics in MOS structures has been found to provide much useful information about the mechanism of certain type of devices. The LB films of Ca-behenate have been investigated with orthophenanthroline with three stearate chains. In these studies, the p-type silicon surfaces (metal-insulator-semiconductor [MIS]), after etching in CP4, cleaning, and drying, were used to deposit LB films, and were covered with aluminum in the form of two circles connected by a channel.

From the linear relation between 1/C and **N**, the thickness was estimated to be 15 Å. A small variation in bias voltage gave rise to a large modulation of AC, which has been ascribed to the small thickness of the insulator and from the high resistivity (ca. 600 Ω cm) of the p-type silicon.

In other studies on MOS structures, the two types of hysteresis, normal and abnormal, are suggested to arise from the ion displacement in the insulator and to the trapping at the interface states. The presence of site-radiation-induced polymerization has been used to provide increased film stability and has been described as an application for high-resolution electron beam lithography for the fabrication of microcircuitry.

So far, only polymerizable sites have been described, and monomolecular layers involving radiation-sites appear unreported. One approach to such a site is the inclusion of a quaternary center and/or a heavy element in the chain to enhance electron beam and x-ray cross section. The transfer rate of the dipalmitoyl lecithin (DPPC) monolayer from aqueous phase to glass surface has been investigated. The change in the contact angle of DPPC monolayer/glass interface, where DPPC was kept at a fixed area/molecule, 134 Å5, was reported to be sigmoidally increased. These rates were analyzed as first-order kinetics, and the transfer rate was found to be 2.59 10^{-3} 1/s. Synthesis and monolayer studies were reported for a geminally branched acid— O,O-dimethyl-arachidic acid—in order to assess the feasibility of forming stabler films in such a system, films that should possess enhanced radiation sensitivity. Surprisingly, the quaternary center did not markedly disturb the formation of monomolecular films, and films formed and transferred to a solid surface were studied by optical and x-ray photoelectron spectroscopic measurements.

The interactions between insoluble monolayers of ionic surfactants and ions dissolved in the subsolution have been the subject of continuing scientific interest since the time of the pioneering investigations of Adam and Langmuir (Adamson and Gast, 1997; Gaines, 1966; Birdi, 1989). This interest rests upon the desire to model important natural and technological processes such as enzymatic activity, permeability of cellular membranes, ion exchange processes in soils, adsorption in ion exchange resins, ion flotation, and chromatography.

A recent report describes the radioactive labeled chromium (iii) ion adsorption on stearic acid LB films. The adsorption of chromium (iii) on a stearic acid monolayer on the surface of $CrCl_3$ was described. Stearic acid monomolecular films on 10^{-3} M $CaCl_2$ subsolutions were deposited in paraffin-coated microscope glass slides by the LB technique (pH range 2–9).

The combination of the LB technique and the neutron activation method of analysis was used to determine the stoichiometry of the interaction between fatty acid (arachidic acid) and metal ions dissolved in the subphase. The experimental data were used to estimate the stability constants of arachidic acid and bivalent metal ions (Cd and Ba). The data were explained as an interaction between metal ions and the monolayer as an adsorption process:

$$2 \text{ RCOO}^- + \text{Me}^{++} \Leftrightarrow (\text{RCOO})^2 \text{ Me} \qquad (4.38)$$

The equilibrium constant is given by the mass action law:

$$\beta = X(R_2Me)/(XR^-)^2 (M_2+(Me)) \tag{4.39}$$

where R– = RCOO-. Furthermore, the following can be derived from these:

$$\beta = [X(R_2Me) X^2(H^+)]/(X^2(RH) X (Me_2+) Kd^2) \tag{4.40}$$

where $K_d = [X(R^-) X(H^+)]/X(RH)$

The foregoing expression represents the Langmuir adsorption isotherms, in which the Me^{2+} ion occupies two adsorption sites, the number of the lattice sites being dependent on the subphase pH. The interactions in the monolayer and the subphase are neglected. The magnitude of X (R_2Me) can be estimated from the experimental known ratio (Me/RCOOH).

The data for various binding constants have been reported in the literature. These data were found to be in good agreement with other literature studies.

4.6.4 PHYSICAL PROPERTIES OF LB FILMS

The surface property of a solid changes even after one layer of lipid is formed as an LB film, such as, the contact angle decreases after LB deposition. Similarly, many other physical methods have shown that LB films change the surface characteristics (Gaines, 1966; Adamson and Gast, 1997; Birdi, 2002a).

Fourier transform infrared (FTIR)-attenuated total reflection (ATR) spectra have been measured for LB films of stearic acid deposited on a germanium plate with 1,2,3,5, and 9 monolayers. In this study, special attention was given to obtaining accurate and unequivocal examinations of changes in structure, subcell packing, and molecular orientation as a function of the number of monolayers deposited. C=O stretching band at 1702 cm^{-1} was missing for the monolayer. The intensity increased linearly for multilayer samples. The CH scissoring band at 1468 cm^{-1} appeared as a singlet in the case of the 1 monolayer. A doublet at 1473 cm^{-1} and 1465 cm^{-1} was observed for films containing more than 3 monolayers. Band progression due to -CH_2- wagging vibration of the trans-zigzag hydrocarbon chains is known to appear between 1400 cm^{-1} and 1180 cm^{-1}. The intensities increased in this region with the number of layers. Monolayers of a surface-active dye (1-methyl-1′-octyadecyl-2,2′ cyanine perchlorate, S-120) incorporated into inert matrix material (e.g., methyl arachidate + arachidic acid) were transferred onto structurally defined silver films by the LB method. The dye-containing monolayers were spaced from the surface by accurately known increments by the deposition of inert spacer monolayers.

Surface-enhanced resonance Raman spectra were observed from dye molecules spaced as distant as six spacer increments (ca. 16 nm = 16 Å) from the silver surface. These studies suggested that an electromagnetic mechanism is operative in this assembly in contradistinction to a chemical mechanism that would require direct contact between the Raman-active species and the metal surface. These studies are of relevance in the study of chromophoric species in biological membranes (e.g., enzymes, redox proteins, and chlorophylls).

In order to compare the monolayer properties at the air–water interface and the spectral characteristics in the visible range, purple membrane films were transferred to glass slides. This was done only on withdrawal across the interface. The slides emerged wet and needed to be dried thoroughly before being dipped again (after ca. 20 min). If the wet slide is dipped again without drying, the adsorbed film will float off the support on the downstroke. This is a general behavior, as also observed for different proteins (e.g., hemoglobin, BSA, ovalbumin, and lipids [lecithin, cholesterol]) (Birdi, 1999). These LB films were prepared with films up to 160 layers. The visible spectra of 80 layers of bacteriorhodopsin/soya-PC showed a single absorption band at λ_{max} at 570 nm, and is identical to the spectrum of an aqueous suspension of the purple membrane. The absorption at 570 nm was found to be linearly proportional to the number of LB layers. These analyses also indicated that all bactriorhodopsin molecules at the interface remain spectroscopically intact.

The LB monolayers of dimethyldioctyadecylammonium ions on molecularly smooth muscovite mica surfaces were investigated. Direct measurements of the interaction between such surfaces were carried out using the surface force apparatus. A long-range attractive force considerably stronger than the expected van der Waals force was measured. Studies on the electrolyte dependence of this force indicate that it does not have an electrostatic origin but that the water molecules were involved in this.

The electrical potential across a LB film of dioleoyl-lecithin deposited onto a fine-pore membrane, imposed between equimolar aqueous solutions of NaCl and KCl, was reported to exhibit rhythmic and sustained pulsing or oscillations of electrical potential between the two solutions. These oscillations were attributed to the change of permeability of Na^+ and K^+ ions across the membrane, which originated from the phase transition of lecithin.

The LB monolayers of cyanine dye have been studied by various methods. A new method was proposed for preparing and removing the J aggregates of some cyanine dyes that did not have long alkyl chains. The cyanine dye and arachidic acid were mixed in a solution, and the film was deposited on a quartz plate by the LB method. The dye molecules were found to be contained in the film, and some films showed a remarkable J band and resonance fluorescence in the longer wavelength region than in the corresponding monomer band.

Catalysis and LB: Catalysis by electrodeposited platinum in an LB film on a glassy carbon electrode were investigated (Birdi, 1999).

4.6.5 APPLICATION OF LB FILMS IN INDUSTRY

The LB film is thus seen to be a sensitive structure in which a layer of lipid can be arranged on any solid. The applications of such structures in various industries have been much exploited recently. Since even a single monolayer of lipid on a solid surface (such as glass, metal, etc.) gives a large change in the contact angle of water, indicating the potential of such a procedure.

Thin films of phthalocyanine compounds in general, and those prepared by the LB method in particular, display novel electrical properties (Baker, 1985). The LB technique for depositing mono- and multilayer coatings with well-controlled thickness and morphology offers excellent compatibility with microelectronic technology. Such films have recently been reviewed for their potential applications. The combination of LB supramolecular films with small dimensionally comparable microelectronic substrates affords new opportunities for generation of fundamental chemical property information and evaluation of new organic thin-film semiconductors as microelectronic components and devices.

4.6.6 BILIPID MEMBRANES (BLMs)

Before the advent of the electron microscope, the structures of biological membranes were not fully understood at the molecular level. However, it was realized that biological cells were contained by some kind of a thin *lipid membrane*. In order to analyze this in more detail, experiments were conducted as follows. Lipids were extracted from biological cells and compressed on the Langmuir balance, and the value of area per molecule was estimated (ca. 45 Å 2/molecule). Knowing the diameter of the cells and from the amounts of lipids (and the area/molecule data), the conclusion was reached that the cell membranes were composed of *bilayer of lipids* (*bilipid membrane or BLM*). This was one of the most important results in the history of biological cell membranes. Later, of course, these results were confirmed from x-ray diffraction data and other scanning probe microscopes (SPMs) (Birdi, 1999, 2002a).

4.7 VESICLES AND LIPOSOMES

It was mentioned that ordinary surfactants (soaps, etc.), when dissolved in water, form self-assembly micellar structures. Phospholipids are molecules like surfactants; they also have a hydrophilic head and generally have two hydrophobic alkyl chains. These molecules are the main components of cell membranes. Lung fluid also consists mainly of lipids of this kind. In fact, usually, cell membrane are made up of two layers of phospholipids, with the tails turned inward, in an attempt to avoid water. The external membrane of a cell contains all the organelles and the cytoplasm.

Phospholipids, when dispersed in water, may exhibit self-assembly properties (either as micellar self-assembly aggregates or larger structures). This may lead to aggregates that are called liposomes or vesicles. Liposomes are structures that are empty cells and that are currently being used by some industries. They are microscopic vesicles or containers formed by the membrane alone, and are widely used in the pharmaceutical and cosmetic fields because it is possible to insert chemicals inside them. Liposomes may also be used solubilize (in its hydrophobic part) hydrophobic chemicals (water-insoluble organic compounds) such as oily substances so that they can be dispersed in an aqueous medium by virtue of the hydrophilic properties of the liposomes (in the alkyl region).

A certain type of lipid (or lipid-like) molecules are found that when dispersed in water tend to make self-assembly structures (Figure 4.13). Detergents were shown to aggregate to spherical or large cylindrical-shaped micelles. It is known that if egg phosphatidylethanolamine (egg lecithin) is dispersed in water at 25°C, it forms a self-assembly structure, which is called liposome or vesicle.

A liposome is a spherical vesicle with a membrane composed of a phospholipid and cholesterol (less than 50%) bilayer. Liposomes can be composed of naturally derived phospholipids with mixed lipid chains (such as egg phosphatidylethanolamine) or

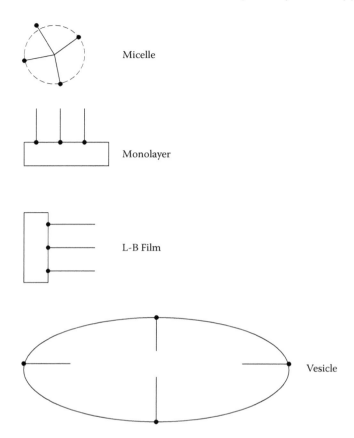

FIGURE 4.13 Different self-assembly (SA) structures: (a) micelle; (b) monolayer; (c) LB film; (d) vesicle; (e) liposome.

pure surfactant components such as DOPE (dioleolylphosphatidylethanolamine). Liposomes usually contain a core of aqueous solution. Vesicles are *unilamellar* phospholipid liposome.

The word *liposome* comprises two terms (from Greek—*lipid* [fat] and *soma* [body]). It does not in itself denote any size characteristics. Furthermore, the term liposome does not necessarily mean that it must contain lipophobic contents, such as water, although it usually does. The vesicles may be conceived as microscopic (or *nano-sized*) containers of carrying molecules (drugs) from one place to another. The structures are suitable for both transporting water-soluble or water-insoluble drugs. Since the lipids used are biocompatible molecules, this may also enhance their adsorption and penetration into the cells.

During the past decades, liposomes have been used for drug delivery on account of their unique solubilization characteristics for water-insoluble organic substances. A liposome encapsulates a region of aqueous solution inside a hydrophobic membrane; dissolved hydrophilic solutes cannot readily pass through the lipids. Hydrophobic chemicals can be dissolved into the membrane, and this way, the liposome can carry both hydrophobic and hydrophilic molecules. To deliver the molecules to the sites of action, the lipid bilayer can fuse with other bilayers such as the cell membrane, thus delivering the liposome contents. By making liposomes in a solution of DNA or drugs (which would normally be unable to diffuse through the membrane), they can be delivered (indiscriminately) past the lipid bilayer.

Liposomes can also be designed to deliver drugs in different specific ways. Liposomes that contain low (or high) pH can be constructed in such a way that dissolved aqueous drugs will be charged in solution (i.e., the pH is outside the drug's pI range) (isoelectric pH = pI). As the pH naturally neutralizes within the liposome (protons can pass through a membrane), the drug will also be neutralized, allowing it to freely pass through a membrane. These liposomes work to deliver drug by diffusion rather than by direct cell fusion. Another strategy for liposome drug delivery is to target endocytosis events. The size range of liposomes can be tailored so that they are suitable for viable targets for natural macrophage phagocytosis. They are then consumed while in the macrophage's phagosome. In this process, the drug is target oriented.

Further, another important property of liposomes is their natural property to target cancer cells. The endothelial wall of all healthy human blood vessels are encapsulated by endothelial cells that are bound together by tight junctions. These tight junctions block large vesicles of in the blood from leaking out of the blood vessel. It is known that tumor vessels do not contain the same level of seal between cells, and they are diagnostically leaky. The size of liposomes can be varied to suit a specific application. For example, liposomes of certain sizes (typically less than 400 nm) can rapidly enter tumor sites from the blood but are kept in the bloodstream by the endothelial wall in healthy tissue vasculature. Liposome-based anticancer drugs (Doxorubicin [Doxil] and Daunorubicin [Daunoxome]) are now being used as drug delivery systems.

Liposomes can be created by shaking or sonicating phospholipids in water. Low shear rates create *multilamellar* liposomes, which have many layers like an onion. Continued high-shear sonication tends to form smaller *unilamellar* liposomes. In this technique, the liposome contents are the same as the contents of the aqueous phase. Sonication is generally considered a "gross" method of preparation, and newer methods such as extrusion are employed to produce materials for human use.

Further advances in liposome research have been able to allow liposomes to avoid detection by the body's immune system. These liposomes are known as *stealth liposomes*, and have been made with polyethylene glycol (PEG) studding the outside of the membrane. The idea behind this is as follows. The PEG coating has been found to exhibit neutral properties in the body. In other words, because of electroneutrality, stealth liposomes do not attach to cells, and can circulate in the system for much longer periods. The latter trait will ensure a longer circulatory life for the drug delivery mechanism. Specific targeting is also used besides the PEG coating in stealth liposomes. Such target-oriented liposomes are employed in delivering drugs to the tissues where they are needed, since some drugs are indeed toxic at high concentrations. This is achieved by attaching a specific biological species to the liposome, which enables it to bind to the targeted drug delivery site. Different targeting ligands have been used (monoclonal antibodies, vitamins, or specific antigens) for this purpose, and are found in the literature. Further, it has been reported that liposomes can be covered with ligands to activate endocytosis in other cell types.

4.7.1 Applications in Drug Delivery

Drugs are applied in various ways to treat illness. A drug designed to cure liver or lung must reach its target with a suitable concentration. The main object is to treat the illness in any particular organ, and the drug dosage is determined accordingly. However, if the drug breaks down in the process of transport through the stomach, etc., then other innovations are needed. In the following text, an example is given where drug delivery is designed through the nasal pathway.

Inhalable drug delivery: At present, there are many drugs that are applied through the nasal pathway (inhalable drug delivery [IDD]). Besides small molecules (such as hormones), even much larger molecules (such as insulin and other proteins) have been reported as useful IDD systems. However, they need to meet certain critical demands:

Deliver the drug effectively and reach the lung
The particles (in the form of aerosols) need to be designed to achieve consistent delivery

Surface-active substances, which are known to enhance penetration through the skin barrier, also needs to be added. These should, of course, not cause any irritation in the nose and other air pathways. Insulin is currently being marketed commercially for IDD.

4.8 APPLICATIONS

4.8.1 Diagnostics (Immobilized Enzymes on Solid Surfaces)

In the past decade, extensive diagnostic instruments have become available to determine the state of illness control. For instance, the concentration of glucose (in the case of diabetes control) in blood can be easily measured today by using a strip (size: 1 mm × 1 mm) covered with a suitable enzyme (glucose oxidase), which, in contact with the blood sample, reacts (within 30 s) to produce degraded substances of glucose (hydrogen peroxide). This enzyme is very specific to the degradation of glucose. The reaction is calibrated to produce an electrical signal, and millimole per liter (from 3 to 30 mmol/L) or milligram per liter glucose from a small drop of blood can be safely measured. The preparation of the diagnostic strip requires an even layer of the enzyme (or any other suitable chemical) on the test strip. It can be controlled using the following surface chemistry principles:

Contact angle
Surface tension of the applied solution
Use of AFM to make image analyses

5 Solid Surfaces— Adsorption

5.1 INTRODUCTION

In a large variety of applications, the *surface* of a solid plays an important role (e.g., active charcoal, talc, cement, sand, catalysis). Solids are rigid structures and resist any stress effects. Many such considerations in the case of solid surfaces will be somewhat different for liquids. The surface chemistry of solids is extensively described in the literature (Adamson and Gast, 1997; Birdi, 2002). Mirror-polished surfaces are widely applied with metals, where the adsorption at the surface is much more important. Further, the *corrosion* of metals initiates at the surfaces, thus requiring treatments based on surface properties. As described in the case of liquid surfaces, analogous analyses of solid surfaces can be carried out. The molecules at the solid surfaces are not under the same force field as in the bulk phase (Figure 5.1).

The differences between *perfect* crystal surfaces and surfaces with *defects* are very obvious in many everyday observations. The solids were the first material that was analyzed at the molecular scale, leading to an earlier understanding of the structures of solid substances and the crystal atomic structure. This is because while molecular structures of solids can be investigated by such methods as x-ray diffraction, the same analyses for liquids are not as straightforward. These analyses have also shown that there exists surface defects at the molecular level. As pointed out for liquids, when the surface area of a solid powder is increased by grinding, then surface energy is needed. Of course, due to the energy differences between the solid and liquid phases, these processes will be many orders of magnitude different from each other. The liquid state, of course, retains some structure that is similar to its solid state, but in the liquid state, molecules exchange places. The average distance between molecules in the liquid state is roughly 10% larger than in its solid state. It is thus desirable at this stage to consider some of the basic properties of liquid–solid interfaces.

The surface tension of a liquid becomes important when it comes into contact with a solid surface. The interfacial forces are responsible for self-assembly formation and stability on solid surfaces. The interfacial forces present between a liquid and solid can be estimated by studying the shape of a drop of liquid placed on any smooth solid surface (Figure 5.2).

The balance of forces as indicated, again, were analyzed very extensively in the last century by Young (1805), who related the different forces at the solid–liquid boundary and the contact angle, θ, as follows (Adamson and Gast, 1997; Chattoraj and Birdi, 1984: Birdi, 1997, 2002):

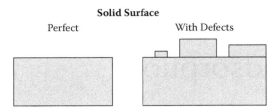

FIGURE 5.1   Solid surface molecules' defects: (a) perfect crystal; (b) surface with defects.

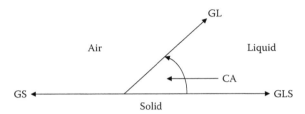

FIGURE 5.2   The state of equilibrium between the surface tensions of liquid (GL)–solid (GS)–liquid/solid (GLS)–contact angle (CA).

$$(\text{Surface tension of solid}) \ (\gamma_S) = \text{Surface tension of solid/liquid} \ (\gamma_{SL}) +$$
$$\text{Surface tension of liquid} \ (\gamma_L) \ (\text{Cos} \ (\theta)) \tag{5.1}$$

$$\gamma_S = \gamma_L \cos(\theta) + \gamma_{SL} \tag{5.2}$$

$$\gamma_L \cos(\theta) = \gamma_S - \gamma_{SL} \tag{5.3}$$

where γ is the interfacial tension at the various boundaries between solid, S, liquid, L , and air (or vapor) phases, respectively. The relation of Young's equation is easy to understand as it follows from simple physics laws. At the equilibrium contact angle, all the relevant surface forces come to a stable state (Figure 5.2).

The geometrical force balance is considered only in the X-Y plane. This assumes that the liquid does not affect the solid surface (in any physical sense). This assumption is safe in most cases. However, only in very special cases, if the solid surface is soft (such as with contact lens), then tangential forces will also need to be included in this equation (as extensively described in the literature). There exists extensive data that convincingly support the equation for liquids and solids.

5.2 SOLID SURFACE TENSION (WETTING PROPERTIES OF SOLID SURFACES)

The wetting of solid surfaces is very apparent when considering the difference between Teflon and metal surfaces. To understand the degree of wetting between the liquid, L, and the solid, S, it is convenient to rewrite Equation 5.3 as follows:

$$\cos(\theta) = (\gamma_S - \gamma_{LS})/\gamma_L \tag{5.4}$$

which would then allow one to analyze the variation of γ with the change in the other terms. The latter is important because complete wetting occurs when there is no finite contact angle, and thus $\gamma_L <> \gamma_S - \gamma_{LS}$. However, when $\gamma_L > \gamma_S - \gamma_{LS}$, then $\cos(\theta) < 1$, and a finite contact angle is present. The latter is the case when water, for instance, is placed on a hydrophobic solid, such as Teflon, polyethylene, or paraffin. The addition of surfactants to water, of course, reduces γ_L, therefore, θ will decrease on the introduction of such surface-active substances.

A complete discussion of wetting of solids is beyond the scope of this book, and the reader is therefore encouraged to look up other standard textbooks on surface chemistry (Adamson and Gast, 1997; Chattoraj and Birdi, 1984; Birdi, 1997, 2002). The state of a fluid drop under dynamic conditions, such as evaporation, becomes more complicated (Birdi et al., 1988). However, in this text, we are interested in the spreading behavior when a drop of one liquid is placed on the surface of another liquid, especially when the two liquids are immiscible. Harkins (1952) analyzed the spreading phenomena by introducing a quantity, spreading coefficient **Sa/b**, defined as (Adamson and Gast, 1997; Birdi, 2002)

$$S_{a/b} = \gamma_a - (\gamma_b + \gamma_{ab}) \qquad (5.5)$$

where $S_{a/b}$ is the spreading coefficient for liquid **b** on liquid a, γ a and γ b are the respective surface tensions, and γ_{ab} is the interfacial tension between the two liquids. If the value of $S_{b/a}$ is positive, spreading will take place spontaneously, whereas, if it is negative, liquid **b** will rest as a lens on liquid **a**.

However, the value of γ_{ab} needs to be considered as the equilibrium value, and therefore if one considers the system at nonequilibrium, then the spreading coefficients would be different. For example, the instantaneous spreading of benzene is observed to give a value of **Sa/b** as 8.9 dyn/cm, and therefore benzene spreads on water. On the other hand, as the water becomes saturated with time, the value of water decreases, and benzene drops tend to form lenses. The short-chain hydrocarbons such as hexane and hexene also have positive initial spreading coefficients, and spread to give thicker films. Longer-chain alkanes, on the other hand, do not spread on water (e.g., the $S_{a/b}$ for C_{16}(hexadecane)/water is -1.3 dyn/cm at $25°C$.

It is also obvious that, since impurities can have very drastic effects on the interfacial tensions in Equation 2.24, the value of $S_{a/b}$ would be expected to vary accordingly (see Table 5.1).

The spreading of a solid substance (e.g., cetyl alcohol $[C_{18}H_{38}OH]$), on the surface of water has been investigated in some detail (Gaines, 1966; Adamson and Gast, 1997; Birdi, 2002a). Generally, however, the detachment of molecules of the amphiphile into the surface film occurs only at the periphery of the crystal in contact with the air–water surface. In this system, the diffusion of amphiphile through the bulk water phase is expected to be negligible because the energy barrier now includes not only the formation of a hole in the solid but also the immersion of the hydrocarbon chain in the water. It is also obvious that the diffusion through the bulk liquid is a rather slow process. Furthermore, the value of **Sa/b** would be very sensitive to such impurities as regard spreading of one liquid upon another.

TABLE 5.1
Calculation of Spreading Coefficients, Sa/b, for Air–Water Interfaces (20°C; a = air; w = water; o = oil)

Oil	$\gamma_{w/a} - \gamma_{o/a} - \gamma_{o/w} = Sa/b$	Conclusion
n-$C_{16}H_{34}$	$72.8 - 30.0 - 52.1 = -0.3$	*Will not spread*
n-Octane	$72.8 - 21.8 - 50.8 = +0.2$	*Will just spread*
n-Octanol	$72.8 - 27.5 - 8.5 = +36.8$	*Will spread*

Another example is that the addition of surfactants (detergents) to a fluid dramatically affects its wetting and spreading properties. Many technologies utilize surfactants for control of wetting properties (Birdi, 1997). The ability of surfactant molecules to control wetting arises from their *self-assembly* at the liquid–vapor, liquid–liquid, solid–liquid, and solid–air interfaces and the resulting changes in the interfacial energies (Birdi, 1997). These interfacial self-assemblies exhibit rich structural detail and variation. The molecular structure of the self-assemblies, and the effects of these structures on wetting or other phenomena, remain topics of extensive scientific and technological interest.

In the case of *oil spills* on the seas, these considerations become very important. The treatment of such pollutant systems requires knowledge of the state of the oil. The thickness of the oil layer will be dependent on the spreading characteristics. The effect on ecology (such as birds and plants) will depend on the spreading characteristics.

Young's equation at liquid$_1$–solid–liquid$_2$ has been investigated in various systems where it has been found that the liquid$_1$–solid–liquid$_2$ surface tensions meet at a given contact angle. For example, the contact angle of water drop on Teflon is 50° in octane (Chattoraj and Birdi, 1984; see Figure 5.3).

In the water–Teflon–octane system, the contact angle, θ, is related to the different surface tensions as follows:

$$\gamma_{s\text{-octane}} = \gamma_{\text{water-s}} + \gamma_{\text{octane-water}} \cos(\theta) \qquad (5.6)$$

$$\cos(\theta) = (\gamma_{s\text{-octane}} - \gamma_{\text{water-s}})/\gamma_{\text{octane-water}} \qquad (5.7)$$

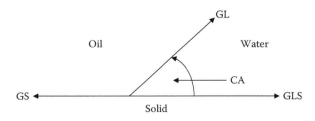

FIGURE 5.3 Contact angle at water–Teflon–octane interface.

This gives the value of $\theta = 50°$ when using the measured values of $\gamma_{s\text{-octane}} - \gamma_{water\text{-}s})/$ $\gamma_{octane\text{-}water}$. The experimental θ is the same. This analysis showed that the assumptions made in derivation of Young's equation are valid.

The most important property of a surface (solid or liquid) is its capability of interacting with other materials (gases, liquids, or solids). All interactions in nature are governed by different kinds of molecular forces (such as *van der Waals, electrostatic, hydrogen bonds, dipole–dipole interactions*). Based on various molecular models, the surface tension, γ_{12}, between two phases with γ_1 and γ_2, was given as (Adamson and Gast, 1997; Chartoraj and Birdi, 1984)

$$\gamma_{12} = \gamma_1 + \gamma_2 - 2\,\Phi_{12}\,(\gamma_1\,\gamma_2)^2 \tag{5.8}$$

where Φ_{12} is related to the interaction forces across the interface. This parameter is easy to expect since the interface forces will depend on the molecular structures of the two phases. In the case of systems such as alkane (or paraffin)–water, there Φ_{12} is found to be unity in the previous equation, if we rewrite the equation as follows: Φ_{12} is unity since the alkane molecule exhibits no hydrogen bonding property, while water molecules are strongly hydrogen bonded.

It is thus found convenient that, in all liquid–solid interfaces, there will be present different (apolar (dispersion) forces + polar (hydrogen-bonding) electrostatic forces). Hence, all liquids and solids will exhibit γ of different kinds as shown

Liquid surface tension:

$$\gamma_L = \gamma_{L,D} + \gamma_{L,P} \tag{5.9}$$

Solid surface tension:

$$\gamma_S = \gamma_{S,D} + \gamma_{S,P} \tag{5.10}$$

This means that γ_S for Teflon arises only from *dispersion* (SD) forces. On the other hand, a glass surface shows γ_S that will be composed of both SD and SP. Hence, the main difference between Teflon and glass surface will arise from the SP component of glass. These criteria have been found to be of importance in the case of the application of adhesives. The adhesive used for glass will need to bind to solid with both polar and apolar forces.

The values of γ_{SD} for different solids as determined from these analyses are as follows:

Solid	γ	γSD	γSP
Teflon	19	19	0
Polypropylene	28	28	0
Polycarbonate	34	28	6
Nylon 6	41	35	6
Polystyrene	35	4	1
PVC	41	39	2

Solid	γ	γSD	γSP
Kevlar49	39	25	14
Graphite	44	43	1

Most of the notions and physical laws of surfaces have been obtained by the studies of liquid–gas or liquid$_1$–liquid$_2$ interfaces. The solid surfaces have been studied in more detail, but this has taken place more recently. The *asymmetrical* forces acting at the surfaces of liquids are much shorter than those expected on solid surfaces. This is due to the high energies that stabilize the solid structures. Therefore, when one examines a solid surface, the surface roughness will need to be considered.

5.2.1 SOLID SURFACE TENSION (γ_{SOLID})

It was described exhaustively before that the molecules at the surface of a liquid are under tension due to asymmetrical forces, which gives rise to surface tension. However, in the case of solid surfaces, one may not envision this kind of asymmetry as clearly, although a simple observation might help one to realize that such surface tension analogy exists. For instance, let us analyze the state of a drop of water (10 μL) as placed on two different smooth solid surfaces (e.g., Teflon and glass). One finds that the contact angles are different (Figure 5.4).

Since the surface tension of water is the same in the two systems, the difference in contact angle can only arise due to the surface tension of solids being different. The surface tension of liquids can be measured directly (as described in Chapter 2). However, this is not possible in the case of solid surfaces. Experiments show that, when a liquid drop is placed on a solid surface, the contact angle, θ, indicates that the molecules interact across the interface. This shows that these data can be used to estimate the surface tension of solids.

5.3 CONTACT ANGLE (θ) OF LIQUIDS ON SOLID SURFACES

As already mentioned, a solid in contact with a liquid leads to interactions related to the surfaces involved. The solid surface is being brought in contact with surface

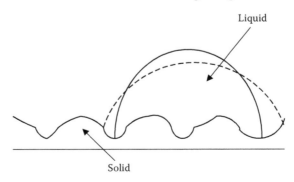

FIGURE 5.4 Apparent or true contact angle of a liquid drop on a rough solid surface.

TABLE 5.2
Contact angles, θ, of Water on
Different Solid Surfaces (25°C)

Solid	θ
Teflon (PTE)	108
Paraffin wax	110
Polyethylene	95
Graphite	86
AgI	70
Polystyrene	65
Glass	30
Mica	10

forces of the liquid (surface tension of liquid). If a small drop of water is placed on a smooth surface of Teflon or glass (Figure 5.4), one finds that these drops are different because there are three surface forces (tensions) that, at equilibrium, give rise to a contact angle θ (Table 5.2).

The relationship Young's equation describes is the interplay of forces (liquid surface tension, solid surface tension, liquid–solid surface tension) at the three-phase boundary line. It is regarded as if these forces interact along a line. Experimental data show that this is indeed true. The magnitude of θ is thus only dependent on the molecules nearest the interface, and is independent of molecules much farther away from the contact line.

Further, it has been defined that

When θ is less than 90°, the surface is wetting (such as water on glass).
When θ is greater than 90°, then the surface is nonwetting (such as water on Teflon).

Most important to mention is that, by treating the glass surface using suitable chemicals, the surface can be rendered hydrophobic. This is the same technology used in many utensils treated with Teflon or similar material.

5.4 MEASUREMENTS OF CONTACT ANGLES

The magnitude of the contact angle, θ, between a liquid and solid can be determined by various methods. The method to be used depends on the system, as well as the accuracy required. The two most common methods are measurement by direct microscope and a goniometer by photography. It should be mentioned that the liquid drop that one generally uses in such measurements is very small, such as 10 to 100 μL. There are two different systems of interest: liquid–solid or $liquid_1$–solid–$liquid_2$. In the case of some industrial systems (such as oil recovery), one needs to determine at high pressures and temperature. In these systems, the value of θ can be measured by using photography. Recently, digital photography has also been used since these data can be analyzed by computer programs.

It is useful to consider some general conclusions derived from these data. One defines a solid surface as *wetting* if θ is less than 90. However, a solid surface is designated as *nonwetting* if θ is greater than 90. This is a practical and semiquantitative procedure. It is also seen that water, due to its hydrogen-bonding properties, exhibits a large θ on nonpolar surfaces (PE, PTE, PE). On the other hand, one finds lower θ values on polar surfaces (glass, mica). However, in some applications, the surface properties may be changed by chemical modifications of the surface. For instance, PS has some weak polar groups at the surface. If one treats the surface with H_2SO_4, it forms sulfonic groups. This leads to values of θ lower than 30 (depending on the time of contact between sulfuric acid and the PS surface). This treatment (or similar) has been used in many applications where the solid surface is modified to achieve a specific property. Since only the surface layer (a few molecules deep) is modified, the bulk of the solid properties do not change. This analysis shows the significant role of studying the contact angle of surfaces in relation to the application characteristics.

The magnitude of contact angle of water (for example) is found to vary depending on the nature of the solid surface. The magnitude of θ is almost 100 on a waxed surface of car paint. The industry strives to create such surfaces to give θ > 150, the so-called superhydrophobic surfaces. The large θ means that water drops do not wet the car polish and are easily blown off by wind. The car polish also is designed to leave a highly *smooth* surface.

In many industrial applications, the concern is for both smooth and rough surfaces. The analyses of θ on *rough surfaces* will be somewhat complicated than those on smooth surfaces. The liquid drop on a rough surface (Figure 5.4) may show the *real* θ (solid line) or some lower value (*apparent*; dotted line), dependent on the orientation of the drop.

However, no matter how rough the surface, the forces will be the same as those that exist between a solid and a liquid. The surface roughness may show contact angle *hysteresis* if one makes the drop move, but this will arise from other parameters (e.g., wetting and dewetting). Further, in practice, the surface roughness is not easy to define. A *fractal* approach has been used to achieve a better understanding (Feder, 1988; Birdi, 1993).

In industry, many systems are found in which contact angle analyses are useful. In Table 5.3, some unusual data for different systems are given.

TABLE 5.3
Contact Angle Data of Different Liquid–Solid Systems

System Liquid	γ	Contact Angle
Liquid Na (100°C)—glass	222	6
Hg (100°C)—glass	460	143

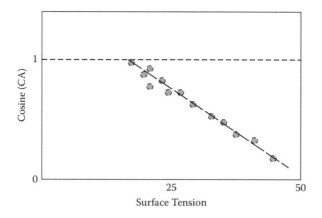

FIGURE 5.5 Plot of contact angle (cos[θ]) of different liquids on Teflon.

Even though Young's equation may sometimes not be easily applicable in all systems, many useful conclusions can be obtained if systematic data are available. In fact, currently, much basic research in industry is based on Young's equation analyses. For example, the data of cos(θ) of various liquids on Teflon gave an almost straight-line plot (Figure 5.5).

These data show that

$$\cos(\theta) = k_1 - k_2 \, \gamma_L \qquad (5.11)$$

This can also be rewritten as

$$\cos(\theta) = 1 - k_3 \, (\gamma_L - \gamma_{cr}) \qquad (5.12)$$

where γ_{cr} is the critical value of γ_L at cos(θ) equal to 0. The values of γ_{cr} have been reported for different solids using this procedure (as suggested by Zisman).

The magnitude of γ_{cr} for Teflon of 18 mN/m thus suggests that $-CF_2-$ groups exhibit this low surface tension. The value of γ_{cr} for $-CH_2-CH_3-$ alkyl chains gave a higher value of 22 mN/m than for Teflon. Indeed, from experience, one also finds that Teflon is a better water-repellant surface than any other material. The magnitudes of γ_{cr} for different surfaces are seen to provide much useful information (Table 5.4).

These data show the surface characteristics as related to γ_{cr}. In many cases, the surface of a solid may not behave as desired, and therefore it treated accordingly, which results in a change of the contact angle of fluids. For instance, the low surface energy of polymers (polyethylene [PE]) is found to change when treated with flame or corona (as shown in the following table).

Material	Liquid	Contact Angle
PE	Water	87
	Corona	55
PE (corona)	Water	66
	Corona	49

TABLE 5.4
Some Typical γ_{cr} Values for Solid Surfaces

Surface Group	γ_{cr}
-CF$_2$-	18
-CH$_2$-CH$_3$-	22
Phenyl	30
Alkyl chlorine	35
Alkyl hydroxyl	40

Source: Birdi, K. S., Ed., *Handbook of Surface and Colloid Chemistry-CD Rom*, CRC Press, Boca Raton; 2002 (Second Edition).
Note: Estimated from Zisman plots.

5.5   ADSORPTION OF GAS ON A SOLID SURFACE

The solid surface interacts with gases or liquids in various phenomena. The adsorption of gas on a solid surface has been known to be of much importance in various systems (especially in industries involved with catalysis). The molecules in gas are moving very fast, but on adsorption (gas molecules are more or less fixed), there will be a large decrease in kinetic energy (thus, a decrease in entropy, ΔS). Adsorption takes place spontaneously,

$$\Delta G_{ad} = \Delta H_{ad} - T \Delta S_{ad} \qquad (5.13)$$

which means that ΔG_{ad} is negative. This suggests that ΔH_{ad} is negative (exothermic).

The adsorption of gas can be of different types. The gas molecule may adsorb as a kind of *condensation* process; it may under other circumstances react with the solid surface (chemical adsorption or chemisorption). In the case of chemiadsorption, a chemical bond formation can almost be expected. On carbon, while oxygen adsorbs (or chemisorbs), one can desorb CO or CO_2. Experimental data can provide information on the type of adsorption. On porous solid surfaces, the adsorption may give rise to *capillary condensation*. This indicates that porous solid surfaces will exhibit some specific properties. Catalytic reactions (e.g., formation of NH_3 from N_2 and H_2) give the most adsorption process in industry.

The surface of a solid may differ in many ways from its bulk composition. Especially, such solids as commercial carbon black may contain minor amounts of impurities (such as aromatics, phenol, carboxylic acid). This would render surface adsorption characteristics different from that of pure carbon. It is therefore essential that, in industrial production, quality control of the surface from different production batch is maintained. Otherwise, the surface properties will affect the application. Another example arises from the behavior of glass powder and its adsorption character for proteins. It has been found that if glass powder is left exposed to the air, its surface may become covered by pollutants. This leads to lower adsorption of proteins than on a clean surface. A silica surface has been considered to exist as O-Si-O as well as hydroxyl groups formed with water molecules. The orientation of the different groups may also be different at the surface.

Carbon black has been reported to possess different kinds of surface chemical groups. These are aromatics, phenol, carboxylic, etc. The different sites can be estimated by comparing the adsorption characteristics of different adsorbents (such as hexane and toluene).

When any *clean* solid surface is exposed to a gas, the latter may adsorb on the solid surface to varying degree. It has been recognized for many decades that gas adsorption on solid surfaces does not stop at a monolayer state. Of course, more than one layer (multilayer) adsorption will take place only if the pressure is reasonably high. Experimental data show this when the volume of gas adsorbed, v_{gas}, is plotted against P_{gas} (Figure 5.6).

These analyses showed that there were five different kinds of adsorption states (Figure 5.6). The adsorption isotherms were classified based on v_{gas} versus P_{gas} data by Brunauer.

Type I: These are obtained for Langmuir adsorption.

Type II: This is the most common type where multilayer surface adsorption is observed.

Type III: This is a somewhat special type with usually only a multilayer formation, such as nitrogen adsorption on ice.

Type IV: If the solid surface is porous, this is similar to type II.

Type V: On porous solid surfaces type III.

The pores in a porous solid surface are found to vary from 2 nm to 50 nm (*micropores*). *Macropores* are designated for surfaces larger than 50 nm. *Mesopores* are used for 2 to 50 nm range.

5.5.1   GAS ADSORPTION MEASUREMENT METHODS

5.5.1.1   Volumetric Change Methods

The change in the volume of gas during adsorption is measured directly in principle, and the apparatus is comparatively simple (Figure 5.7).

A mercury reservoir beneath the manometer and the burette are used to control the levels of mercury in the apparatus in Figure 5.7. Calibration involves measuring the volumes of the gas (v_g) lines and the void space (Figure 5.7). All pressure measurements are made with the right arm of the manometer set at a fixed zero point so that the volume of the gas lines does not change when the pressure changes. The apparatus, including the sample, is evacuated, and the sample is heated to remove any previously adsorbed gas. A gas such as helium is usually used for the calibration since it exhibits very low adsorption on the solid surface. After helium is pushed into the apparatus, a change in volume is used to calibrate the apparatus, and the corresponding change in pressure is measured.

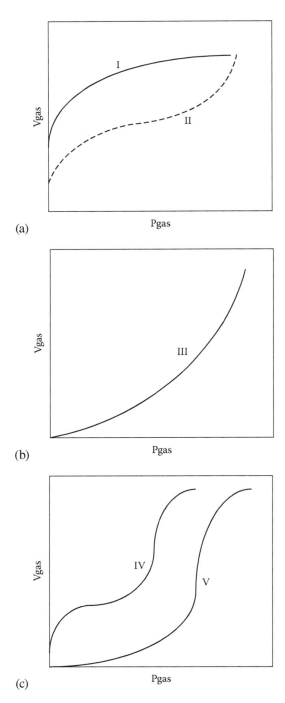

FIGURE 5.6   Plot of (relative volume) v_{gas} versus (relative pressure) P_{gas} (schematic) (types (a) I and II, (b) III, and (c) IV and V).

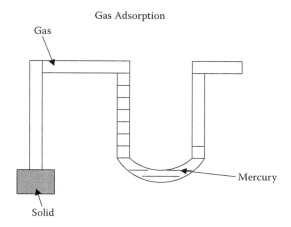

FIGURE 5.7   Gas adsorption apparatus.

Different gases (such as nitrogen) are normally used as the adsorbate if the surface area of a solid needs to be examined. The gas is cooled by liquid nitrogen. The tap to the sample bulb is opened, and the drop in pressure is determined. In the surface area calculations, a value of 0.162 nm^2 is used for the area of an adsorbed nitrogen molecule.

Because of the toxic properties of Hg, modern apparatus use a combination of valves to measure the change in the volume of gas adsorbed. Commercially available instruments are designed with such modern detectors.

5.5.1.2   Gravimetric Gas Adsorption Methods

The amount of gas adsorbed on the solid surface is very small. A modern sensitive microbalance is used to measure the adsorption isotherm. Its sensitivity is very high since only the difference in weight change is measured. These microbalances can measure weight differences in the range of nanograms to milligrams. With such extreme sensitivity, it is possible to measure the weight change caused by the adsorption of a single monolayer on a solid if the surface area is large. The normal procedure is to expose the sample to the adsorbate gas at a certain pressure, allowing sufficient time for equilibrium to be reached, and then determining the mass change. This is repeated for a number of different pressures, and the number of moles adsorbed as a function of pressure is plotted to give an adsorption isotherm.

Microbalances (stainless steel) can be made to handle pressures as high as 120 Mpa (120 atm) since gases that adsorb weakly or boil at very low pressures can still be used.

5.5.1.3   Langmuir Gas Adsorption

This equation assumes that only *one layer* of gas molecule adsorbs. A monolayer of gas adsorbs in the case in which there are only a given number of adsorption sites for a monolayer. This is the most simple adsorption model. The amount adsorbed, N_s, is related to the monolayer coverage, N_{sm}, as follows (Appendix C):

$$N_s/N_{sm} = \mathbf{a}\, p/(1 + \mathbf{a}\, p) \qquad (5.14)$$

where p is the pressure and $\mathbf{a}$ is dependent on the energy of adsorption. This equation can be rearranged as

$$p/N_s = (1/[\mathbf{a}\, N_{sm}] + p/N_{sm}) \qquad (5.15)$$

From the experimental data, p/N_s versus p can be plotted. The plot will be linear, and the slope is equal to $1/N_{sm}$. The intersection gives $\mathbf{a}$. Charcoal is found to adsorb 15 mg of N_2 as monolayer. Another example is that of adsorption of N_2 on mica surface (at 90 K). The following data were found:

Pressure/Pa	Volume of Gas Adsorbed (at STP)
0.3	12
0.5	17
1.0	24

In this equation one assumes that

The molecules adsorb on definite sites.
The adsorbed molecules are stable after adsorption.

The surface area of the solid can be estimated from the plot of p/N_s versus p. Most data fit this equation under normal conditions and are therefore widely applied to analyze the adsorption process.

Langmuir adsorption is found for the data of nitrogen on mica (at 90 K). The data were found to be

$$p = 1/\text{Pa} \qquad 2/\text{Pa}$$

$$\mathbf{V}s = 24 \text{ mm}^3 \quad 28 \text{ mm}^3$$

This shows that the amount of gas adsorbed increases by a factor $28/24 = 1.2$ when the gas pressure increases twofold.

5.5.1.4 Various Gas Adsorption Equations

Other isotherm equations begin as an alternative approach to the developed equation of state for a two-dimensional ideal gas. As mentioned earlier, the ideal equation of state is found to be

$$\Pi\, \mathbf{A} = k_B\, T \qquad (34.1)$$

In combination with the Langmuir equation, one can derive the following relation between N_s and p:

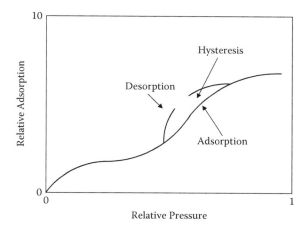

FIGURE 5.8 Adsorption (Ns/Nsm = relative adsorption) versus pressure (*p/po* = relative pressure) of a gas on solid.

$$N_s = K\,p \tag{5.16}$$

where **K** is a constant. This is the well-known Henry's law relation, and it is found to be valid for most isotherms at low relative pressures. In those situations where the ideal Equation 5.16 does not fit the data, the van der Waals equation type of corrections have been suggested.

The adsorption–desorption process is of interest in many systems (such as with cement). The water vapor may condense in the pores after adsorption under certain conditions. This may be studied by analyzing the adsorption–desorption data (Figure 5.8).

5.5.1.5 Multilayer Gas Adsorption

In some systems, adsorption of gas molecules proceeds to higher levels where multilayers are observed. From data analyses, one finds that multilayer adsorption takes place (Figure 5.9).

The BET equation has been derived for multilayer adsorption data.

The enthalpy involved in multilayers is related to the differences, and was defined by BET theory as

$$E_{BET} = \exp[(E_1 - Ev)/RT] \tag{5.17}$$

where E_1 and Ev are enthalpies of desorption. The BET equation, after modification of the Langmuir equation, becomes

$$p/(N_s(p^\circ - p)) = 1/E_{BET}\,N_{sm} + [(E_{BET} - 1)/(E_{BET}\,N_{sm})](p/po) \tag{5.18}$$

A plot of adsorption data of the left-hand side of this equation versus relative pressure (p/po) allows one to estimate N_{sm} and E_{BET}. The magnitude of E_{BET} is found to give

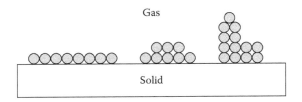

FIGURE 5.9   Brunuer–Emmett–Teller (BET) model for multilayer adsorption on solids.

either data plots that are as type III or II. If the value of E_{BET} is low, which means the interaction between adsorbate and solid are weak, then type III plots are observed. This has been explained as arising from the fact that if E1# Ev, then molecules will tend to form multilayer in patches rather than adsorb on the naked surface.

If strong interaction exists between the gas molecule and the solid, E1 ≥ Ev, then type II plots are observed. The monolayer coverage is clearly observed at low values of p/po.

5.6   ADSORPTION FROM SOLUTION ON SOLID SURFACES

A clean, solid surface is actually an active center for adsorption from the surroundings (e.g., air or liquid). A perfectly cleaned metal surface, when exposed to air, will adsorb a single layer of oxygen or nitrogen (or water). Or, when a completely dry glass surface is exposed to air (with some moisture), the surface will adsorb a monolayer of water. In other words, the solid surface is not as inert as it may seem to the naked eye. This has many consequences in industry, such as with corrosion control. Accordingly, solid surfaces should always be exposed to vacuum prior to any kind of adsorption studies.

5.6.1   Thermodynamics of Adsorption

Activated charcoal or carbon (with surface area of over 1000 m²/g) is widely used for vapor adsorption and in the removal of organic solutes from water. These materials are used in industrial processes to purify drinking water and swimming pool water, to decolorize sugar solutions as well as other foods, and to extract organic solvents (especially trace amounts of dangerous substances). They are also used as a first oral treatment in hospitals for cases of poisoning. Activated charcoal can be made by heat degradation and partial oxidation of almost any carbonaceous matter of animal, vegetable, or mineral origin. For convenience and economic reasons, it is usually produced from bones, wood, lignite, or coconut shells.

The complex three-dimensional structure of these materials is determined by their carbon-based polymers (such as cellulose and lignin), and it is this backbone that gives the final carbon structure after thermal degradation. These materials, therefore, produce a very porous high-surface-area carbon solid. In addition, the carbon has to be activated so that it will interact with and physisorb (i.e., adsorb physically, without forming a chemical bond) a wide range of compounds. This activation process involves controlled oxidation of the surface to produce polar sites.

5.6.1.1 Adsorption of Acetic Acid onto Activated Charcoal

Adsorption at liquid surfaces can be monitored using the Gibbs adsorption isotherm since the surface energy, γ, of a solution can be readily measured. However, for solid substrates, this is not the case, and the adsorption density has to be measured in some other manner. In the present case, the concentration of adsorbate in solution will be monitored. In place of the Gibbs equation, we can use a simple adsorption model based on the mass action approach.

On any solid surface, there are a certain number of possible adsorption sites per gram (N_m), where any adsorbate can freely adsorb. There will be a fraction θ of sites which are filled by one adsorbing solute. There will also exist an *adsorption–desorption* process at the surface. The rate of adsorption will be given as

$$[\text{concentration of solute}][1- \theta] \, N_m \qquad (5.19)$$

and the rate of desorption will be given as

$$[\text{concentration of solute}][\theta] \, N_m \qquad (5.20)$$

It is known that at equilibrium these rates must be equal:

$$k_{ads} \, C_{bulk} \, (1 - \theta) \, N_m = k_{des} \, \theta \, N_m \qquad (5.21)$$

where k_{ads}, k_{des} are the respective proportionality constants, and C_{bulk} is the bulk solution concentration of the solute. The equilibrium constant, $K_{eq} = k_{ads}/k_{des}$, gives

$$C_{bulk}/\theta = C_{bulk} + 1/K_{eq} \qquad (5.22)$$

and since $\theta = N/N_m$, where N is the number of solute molecules adsorbed per gram of solid, one can write

$$C_{bulk}/N = C_{bulk}/N_m + 1/(K_{eq} \, N_m) \qquad (5.23)$$

Thus, measurement of N for a range of concentrations (C) should give a linear plot of C_{bulk}/N against C_{bulk}, where the slope gives the value of Nm, and the intercept gives the value of the equilibrium constant K_{eq}. This model of adsorption was suggested by Irving Langmuir and is referred to as the *Langmuir adsorption isotherm*. The aim of this experiment is to test the validity of this isotherm equation and to measure the surface area per gram of charcoal, which can easily be obtained from the measured N_m value if the area per solute molecule is known.

Adsorption experiments are carried out as follows. The solid sample (for example, activated charcoal) is shaken in contact with a solution with a known concentration of acetic acid. After equilibrium is reached (after 24 h), the amount of acetic acid adsorbed is determined.

The concentration of acetic acid can be determined by titration with NaOH solution. Solutions of dyes (such as methylene blue) may also be used, and after adsorption, the amount of dye in solution is measured by any convenient spectroscopic method (VIS or UV or fluorescence).

5.6.2 SOLID SURFACE AREA DETERMINATION

In all applications where finely divided powders are used (such as talcum, cement, charcoal powder), the property of these will depend mainly on the surface area per gram (varying from a few square meters [talcum] to over $1000 \, m^2/g$ [charcoal]). For example, if one needs to use charcoal to remove some chemical (such as coloring substances) from wastewater, then it is necessary to know the amount of absorbent needed to fulfill the process. In other words, if a $1000 \, m^2$ area is needed for adsorption when using charcoal, then 1 g of solid will be required. In fact, under normal conditions, swallowing charcoal would be considered dangerous because it would lead to the removal of essential substances from the stomach lining (such as lipids and proteins).

The estimation of the surface area of finely divided solid particles from solution adsorption studies is subject to many of the same considerations as in the case of gas adsorption, but with the added complication that larger molecules are involved whose surface orientation and pore penetrability may be uncertain. A first condition is that a definite adsorption model is obeyed, which in practice means that area determination data are valid within the simple Langmuir Equation 5.23 relation. The constant rate is found, for example, from a plot of the data, according to Equation 5.23, and the specific surface area then follows from Equations 5.21 and 5.22. The surface area of the adsorbent is generally found easily in the literature.

In the case of gas adsorption where the BET method is used, it is reasonable to select the van der Waals area of the adsorbate molecule; moreover, being small or even monoatomic, surface orientation is not a major problem. In the case of adsorption from solution, however, the adsorption may be chemisorption.

In the literature, fatty acid adsorption has been used for surface area estimation since fatty acids are known to pack perpendicular to the surface (self-assembly monolayer formation) and without the close-packed area per molecule of $20.5 \, Å^2$. This seems to be true for adsorption on such diverse solids as carbon black and not too electropositive m and for TiO_2. In all of these cases, the adsorption is probably chemisorption in involving hydrogen bonding or actual salt formation with surface oxygen. Even polar solvents are used to avoid multilayer formation on top of the first layer, the apparent area obtained may vary with the solvent used. In the case of stearic acid on a graphitized carbon surface, Graphon, the adsorption, while obeying the Langmuir equation, appears to be physical, with the molecules flat on the surface.

In another example, the adsorption of surfactants on polycarbonate indicated that, depending on the surfactant and concentration, the adsorbed molecules might be lying flat on the surface perpendicular to it, or might form a bilayer.

A second widely used class of adsorbates is that of dyes. Methods using these are appealing because of the ease with which analysis may be made colorimetrically. The adsorption generally follows the Langmuir equation. Graham found an apparent molecular area of $19.7 \, Å^2$ for methane blue on Graphon or larger than the actual molecular area of 17.5 Å^2, but apparent value for the more oxidized surface of Spheron was about 10.5 Å^2 per molecule. However, this procedure assumes that the dye molecules are present as monomers in the solution. The fatty acid adsorption method has been used by many investigators. Pyridine adsorption has been used on various oxides to obtain surface areas. In the adsorption data that followed the Langmuir equation, the effective molecular area of pyridine is about 24 Å^2 per molecule.

In the literature, many different approaches have been proposed for estimating the surface area of a solid. Surface areas may be estimated from the exclusion of like charged ions from a charged interface. This method is intriguing in that no estimation of either site or molecular area is needed. In general, however, surface area determination by means of solution adsorption studies, while convenient experimentally, may not provide the most correct information. Nonetheless, if a solution adsorption procedure has been standardized for a given system by means of independent checks, it can be very useful determining relative areas of a series of similar materials. In all cases, it is also more real as it is what happens in real life.

5.6.2.1 Experimental Method

Typical procedure used was to take 1.0 g of alumina (as an example) powder and add 10 mL of solution of detergent with varying concentration. The mixture was shaken, and the concentration of detergent was estimated by a suitable method. It was found that equilibrium was obtained after 2–4 h. The detergent such as dodecylammonium chloride was found to adsorb 0.433 mM/g of alumina with a surface area of 55 m²/g. The surface area of alumina, as determined from stearic acid adsorption (and using the area/molecule of 21 Å^2 from monolayer), gave this value of 55 m²/g. These data can be analyzed in more detail:

$$\text{Surface area} = 55 \text{ m}^2/\text{g}$$

$$\text{Amount adsorbed} = 0.433 \text{ mM/g}$$

$$= 0.433 \ 10^{-3} \text{ M } 6 \ 10^{23} \text{ molecules}$$

$$= 0.433 \ 10^{20} \text{ molecules}$$

$$\text{Area/molecule} = 55 \ 10^4 \text{ cm}^2 \ 10^{16} \ \text{Å}^2/ \ 0.433 \ 10^{20} \text{ molecule}$$

$$= 55/0.433 \ \text{Å}^2$$

$$= 127 \ \text{Å}^2$$

The adsorption isotherms obtained for various detergents showed a characteristic feature that an equilibrium value was obtained when the concentration of detergent was over critical micelle concentration (CMC). The adsorption was higher at 40°C than at 20°C. However, the shapes of the adsorption curves was the same (Birdi, 2002).

One can also calculate the amount of a small molecule, such as pyridine (mol. wt. 100), adsorbed as a monolayer on charcoal with 1000 m²/g. In the following, these data are delineated:

Area per pyridine molecule = 24 $\mathring{A}^2$ = 24 10^{-16} cm^2
Surface area of 1 gm charcoal = 1000 m^2 = 1000 10^4 cm^2
Molecules pyridine adsorbed = 1000 10^4 cm^2/24 10^{-16} cm^2/molecule
   = 40 10^{20} molecules
Amount of pyridine per gram charcoal adsorbed =
   (40 10^{20} molecules/6 10^{23})100
   = 0.7 g

This is a useful example to illustrate the application of charcoal (or similar substances with large surface area per gram) in the removal of contaminants by adsorption.

5.6.2.2 Adsorption in Binary Liquid Systems

This discussion so far has been confined to systems in which the solute species are dilute, so that adsorption was not accompanied by any significant change in the activity of the solvent. In the case of adsorption from binary liquid mixtures, where the complete range of concentration, from pure liquid A to pure liquid B, is available, a more elaborate analysis is needed. The terms *solute* and *solvent* are no longer meaningful, but it is nonetheless convenient to cast the equations around one of the components, arbitrarily designated here as component 2.

5.6.3 HEATS OF ADSORPTION (DIFFERENT SUBSTANCES) ON SOLID SURFACES

A solid surface interacts with its surrounding molecules (in the gas or liquid phase) in varying degrees. For example, if a solid is immersed in a liquid, the interaction between the two bodies will be of interest. The interaction of a substance with a solid surface can be studied by measuring the heat of adsorption (besides other methods). The information one needs is whether the process is exothermic (heat is produced) or endothermic (heat is absorbed). This leads to the understanding of the mechanism of adsorption and helps in the application and design of the system. Calorimetric measurements have provided much useful information. When a solid is immersed in a liquid (Figure 5.10), in most cases there is a liberation of heat:

$$q_{imm} = E_S - E_{SL} \qquad (5.24)$$

where E_S and E_{SL} are the surface energy of the solid surface and the solid surface in liquid, respectively.

The quantity q_{imm} is measured from calorimetry where temperature change is measured after a solid (in a finely divided state) is immersed in a given liquid. Since these measurements can be carried out with microcalorimeter sensitivity, there is much systematic data in the literature on this subject. When a polar solid surface is immersed in a polar liquid, it can be expected that there will be a larger q_{imm} than if the liquid was an alkane (nonpolar). The values of some typical systems is depicted in Table 5.5.

These data further show that such studies are sensitive to the surface purity of solids. For example, if the surface of glass powder is contaminated with nonpolar gas, then its q_{imm} value will be lower than that in the case of pure glass.

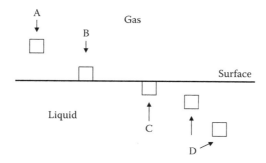

FIGURE 5.10   The process of immersion of a solid in a liquid.

TABLE 5.5
Heat of Immersion (q_{imm}) (erg/cm² at 25°C)

	Liquid	
Solid	Polar (H_2O-C_2H_5OH)	Nonpolar (C_6H_{14})
Polar		
(TiO_2; Al_2O_3; glass)	400–600	100
Nonpolar		
(Graphon; Teflon)	6–30	50–100

5.6.3.1   Solid Surface Roughness (Degree of Roughness)

The nature of the solid surface plays an important role in many applications. In fact, in many applications, it is the main criteria. For instance, friction decreases appreciably as the surface of a solid becomes smooth. This happens because the number of surface molecules that are able to come into contact with another solid or liquid phase are reduced (Figure 5.11).

Thus, in some cases, one prefers roughness (high friction; roads, shoe sole), while in other systems (glass, office table) one requires smooth solid-surface characteristics.

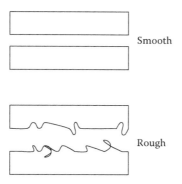

FIGURE.5.11   Profile of a solid rough and smooth surface (schematic).

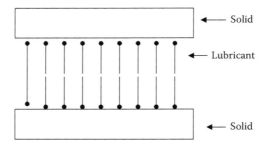

FIGURE 5.12 Friction reduction due to lubrication by adsorbed film (bilayer) of amphiphiles (as lubricant).

The solid surfaces that one finds are manufactured by different methods: sawn, cut, turned, polished, or chemically treated. All of these procedures leave the solid surface rough, in varying degrees. In industry, one finds many methods that can create the roughness.

Polishing is also an important application area of the surface chemistry of solids. The surface layer produced after polishing may or may not remain stable after exposure to its surroundings (air, other gases, oxidation). The polishing industry is much dependent on surface molecular behavior.

5.6.3.2 Friction (between Solid1–Solid2)

Friction is defined as the resistance to sliding between two bodies (Figure 5.12). Friction is lowered using a suitable lubricant (specific for each system; Biresaw and Mittal, 2008). In cases where the solids are very close, the surface roughness becomes the determining factor. That means the resistance is higher between two rough surfaces. The degree of plasticity or deformation of the solids will also effect the friction. Further, the lubricant will give high sliding resistance if its viscosity is high. Thus, when one solid is sliding or rubbing against another, there are a variety of parameters. These parameters are also called "tribology" (related to rubbing).

The *magnitude of friction* can be appreciably reduced if boundary lubrication decreases the force field. This may be achieved by adsorbed films (Figure 5.12). However, one may expect that this is a highly idealized model, and some modifications will be necessary.

5.7 SURFACE TENSION OF SOLID POLYMERS

The surface tension of polymers (synthetic polymers such as plastics, biopolymers such as proteins and gelatin) is indeed of much interest in many areas. In industry where plastics are used, the adhesion of these materials to other materials (such as steel, glass) is of much interest. The adhesion process is very complex since the demand on quality and control is very high. This is also because adhesion systems are part of many life-sustaining processes (such as implants, etc.). The forces involved in adhesion need to be examined, and we will consider some typical examples in the following text.

5.8 APPLICATIONS

5.8.1 FLOTATION OF SOLID PARTICLES TO LIQUID SURFACE

In only rare cases does one find minerals or metals in pure form (such as gold). The earth's surface consists of a variety of minerals (major components: iron, silica oxides, calcium, magnesium, aluminum, chromium, cobalt, and titanium).

Minerals as found in nature are always mixed together (e.g., zinc sulfide and feldspar minerals). In order to separate zinc sulfide, the mixture is suspended in water, and air bubbles are made to achieve separation. This process is called *flotation*, where an ore heavier than water is floated by bubbles.

Flotation is a technical process in which suspended particles are clarified by allowing them to float to the surface of the liquid medium. The material can thus be removed by skimming at the surface. This is economically much cheaper than any other process. If the suspended particles are heavier than the liquid (such as minerals), gas (air, CO_2, or other suitable gas) bubbles can enhance the flotation.

Froth flotation commences by comminution, which is used to increase the surface area of the ore for subsequent processing and to break the rocks into the desired mineral and gangue (which then have to be separated from the desired mineral); the ore is ground into a fine powder. The desired mineral is rendered hydrophobic by the addition of a surfactant or collector chemical; the particular chemical depends on the mineral being refined. As an example, pine oil is used to extract copper. This slurry (more properly called the *pulp*) of hydrophobic mineral-bearing ore and hydrophilic gangue is then introduced to a water bath that is aerated, creating bubbles. The hydrophobic grains of mineral-bearing ore escape the water by attaching to the air bubbles, which rises to the surface, forming a foam (more properly called a *froth*). The *froth* is removed, and the concentrated mineral is further refined.

The flotation industry is a very important area in metallurgy and other related processes. The flotation method is based on treating a suspension of minerals (ranging in size from 10 µm to 50 µm) in the water phase to air (or some other gas) bubbles (Figure 5.13).

Flotation leads to separation of ores from the mixtures. It has been suggested that, among other surface forces, the contact angle plays an important role. The gas (air or other gas) bubble as attached to the solid particle should have a large contact angle for separation (Figure 5.14). Further, it should be stable at the surface.

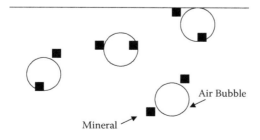

FIGURE 5.13 Flotation of mineral particles as aided by air bubbles.

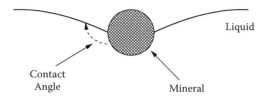

FIGURE 5.14   Contact angle at a mineral particle and liquid interface.

Bubbles as needed for floatation are created by various methods. These may be by

Air injection
Electrolytic methods
Vacuum activation

In a laboratory experiment (Adamson and Gast, 1997), the following recipe may be used. To a 1% sodium bicarbonate solution, add a few grams of sand. Further, some acetic acid (or vinegar) will cause the bubbles of CO_2 produced to cling to the sand particles and thus make these float to the surface. It must be mentioned that, in wastewater treatment, the flotation method is one of the most important procedures.

When rocks in a crushed state are dispersed in water with suitable surfactants (also called *collectors* in industry) to give stable bubbles an aeration, hydrophobic minerals will be floated to the surface by the attachment of bubbles, while the hydrophilic mineral particles will settle to the bottom. The preferential adsorption of the collector molecules on a mineral makes it hydrophobic.

Xanthates have been used for flotation of lead and copper. In these examples, it is the adsorption of xanthane that dominates the flotation.

5.8.2   POLISHING OF SOLID SURFACES

The reflection of light from a smooth solid surface determines the degree of polish. The industrial applications are many, considering the impact of design and appearance on both sales and the product (cars, household appliances, furniture). A very sensitive procedure for the determination of surface roughness has been to use atomic force microscope (AFM; Birdi, 2003).

5.8.3   POWDERS (SURFACE TECHNOLOGY)

The solid material when transformed into fine powder creates a new material (regarding its surface properties and applications). The technology related to this subject is very expansive (found in talcum powder, cement, the clay industry, etc.). The Following areas of powder science are currently being investigated:

Formation and synthesis of particles by different procedures (e.g., grinding, crushing, precipitation, etc.)
Modification of particles by agglomeration, coating, etc.

Characterization of the size, shape, surface area, pore structure, and strength of particles

Origins and effects of interparticle forces

Packing, failure, flow, and permeability of assemblies of particle

Particle–particle interactions and suspension rheology

Handling and processing operations (e.g., slurry flow, fluidization, pneumatic conveying)

Applications of particle technology in production of pharmaceuticals, chemicals, foods, pigments, structural and functional materials, and in environmental and energy-related areas

In a recent report, the effect of synthesis parameters on the precipitation of nanocrystalline boehmite from aluminate solutions has been investigated. Nanocrystalline boehmite ($AlOOH$) is a useful and effective material for the production of Al_2O_3, as applied in many industrial applications as catalyst or catalyst support, membranes, and adsorbents. The preparation conditions applied in the production step of nanocrystalline boehmite strongly affect its morphology, which in turn is reflected in the final transition alumina. In this work, a precipitation method for the production of nanocrystalline boehmite is described, studying the effect of pH, temperature, and aging time on the morphology of the final precipitate. The experiments were performed at temperatures of 30, 60, and 90°C under moderate pH (from 5 to 7) conditions and 1 week of aging in the mother liquor. It is noteworthy that, in these experiments, a unique starting solution is used, and also, the mixing procedure is unique. The starting solution used is a supersaturated sodium aluminate solution. On the other hand, the mixing procedure does not follow the normal route of addition of the neutralization agent (acid) to the aluminate solution. Amorphous boehmite was prepared at 30°C and pH 7 under prolonged aging conditions. At 60°C, pure nanocrystalline boehmite with crystallites 3–8 nm was formed at pH 6 and pH 7 after aging in the mother liquor, while at the higher temperature of 90°C, the formation of pure nanocrystalline boehmite with crystallite size between 3 and 13 nm was obtained at pH 5, pH 6, and pH 7. Aging and temperature influenced the crystallinity of the precipitated phases, with prolonged aging and high temperatures inducing high crystallinity. The pH conditions also had a strong effect on the crystallite size of the precipitates. Actually, for the same temperature and aging time, the higher the pH, the larger the crystallites of the precipitates.

Nanocrystalline boehmite can be synthesized by neutralization of AlOOH liquor under atmospheric conditions (60°C/pH 6), proper aging, and a modification of the usual neutralization procedure.

Other applications of nanoparticles: At present, there are many research reports related to the development and processing of inorganic particles ranging from the *nanoscale* of less than 10 nm to the microscale. The processing comprises various stages of materials fabrication from particles (powders), starting with the particle synthesis to the point of forming and densification to reach the final product.

Some of these projects are related to the development and functionalization of nanoparticles for drug and gene delivery, self-assembling of nanoparticles to achieve unique surface structures with controlled porosity and biofunctionality, and the creation of three-dimensional building blocks for further processing, including sintered products. The main aim has been understanding at a fundamental level these various steps with a view to improving the processing routes for materials of technological importance and to contribute in the development of new and intelligent materials. The high quality of particles in regard to narrow size distribution, form factors, functions, etc., is guaranteed by the high level of characterization methods for particles in the nano and micro range.

5.8.4 Coal Slurry

The need for fuel and energy for mankind is becoming increasingly critical, as far as resources and dependability is concerned. The largest known energy resource is coal, with supplies predicted to last for over 1000 years. The predicted oil and gas supply is in the range of less than 100 years at the current rate of use of approximately 90 million barrels per day.

The coal reserves are very large, but the mobility of coal is difficult (by truck, train, or shiploads) as compared to oil and gas (by pipelines). In order to transfer coal in pipelines, one has to make *coal slurries*. Coal is finely divided, and after that, it is dispersed in water or oil such that a suitable slurry for pipeline transfer can be suitably achieved. Something similar is already being developed in the shale oil industry (in Canada). Coal can thus be transported through pipelines after being dispersed in conjunction with water, oil, etc. This industrial development is based on many aspects related to the surface properties of coal.

In the past decade, a very large interest in this technology has developed. Research is based on studying the surface properties of coal and its adsorption characteristics. Oil production from shales is increasing as the oil prices increase. Additionally, in many coal operations, the dust is controlled by the washwater process.

5.8.5 Earthquakes and Faults (Solid Surfaces)

The occurrence of earthquakes is a highly studied phenomena by geologists. The role of solid surfaces in such phenomena is obvious. Especially, faults are known to contribute to many earthquakes. Faults are treated as shear cracks, the propagation of which may be understood through the application of fracture mechanics. The stability of any fault movement, which determines whether the faulting is seismic or aseismic, is determined by the frictional constitutive law of the fault surface. It is well established that, once a fault has been formed, its further motion is controlled by friction (between the solid surfaces), which arises from contact forces across the two solid surfaces.

6 Wetting, Adsorption, and Cleaning Processes

6.1 INTRODUCTION

When a liquid comes into contact with a solid, there are a few processes of interest to analyze. These are

Wetting
Adsorption
Cleaning

One finds that the *wetting* characteristics of any solid surface play an important role in all kinds of systems. The next most important step is the process of *adsorption* of substances on solid surfaces. These phenomena are the crucial steps for all kinds of *cleaning processes.* Typical systems involved are washing, coatings, adhesion, lubrication, oil recovery, etc.

The liquid–solid or liquid$_1$–solid–liquid$_2$ system is both a contact angle (Young's equation) and capillary phenomena (Laplace equation). These two parameters are

$$Cos\ (\theta) = (\gamma_S - \gamma_{SL})/\gamma_L \qquad (5.4)$$

and

$$\Delta P = (2\ \gamma_L\ Cos\ [\theta])/radius \qquad (6.1)$$

In the following, we will consider some significant phenomena where these parameters are of importance.

6.2 OIL RECOVERY AND SURFACE FORCES

Oil is normally found under high temperatures (80°C) and pressures (200 atm). Oil in reservoirs is produced under high-pressure conditions. The pressure needed depends primarily on the porosity of the reservoir rock and the viscosity of the oil, among other factors. This creates flow of oil through the rocks, which consists of pores of varying sizes and shapes. Roughly, one may compare oil flow to the squeezing of water out of a sponge.

Oil recovery from reservoirs is not a 100% process from rock since not all of the oil in a reservoir is recovered. This means that all the oil recovered until now leaves some 20–40% residual oil in the depleted reservoirs. This may be considered as an advantage in the long run because, as the shortage of oil supplies approaches, one may be forced to develop technologies to recover the residual oil.

In order to enhance this recovery, physicochemical methods have been used. At present, approximately 100 million barrels of oil are used per day. In most oil reservoirs, the primary recovery is based on the natural flow of oil under the gas pressure of the reservoir. In these reservoirs, as this gas pressure drops, a water-flooding procedure is used. The pressure needed is determined by the capillary pressure of the reservoir and the viscosity of the oil. This procedure still results in 30 to 50% of the original oil remaining in the formation. In some cases, such substances as detergents or similar chemicals may be added to enhance the flow of oil through the porous rock structure. The principle is to reduce the Laplace pressure (i.e., $\Delta P = \gamma$/curvature) and to reduce the contact angle as well. This process is called *tertiary oil recovery*. The aim is to retrieve oil that is trapped in capillary-like structures in the porous oil-bearing material. The addition of a surface-active agent (SAA) reduces the oil–water interfacial tension (from ca. 30–50 mN/m to less than 10 mN/m).

In the tertiary process, more complicated chemical additives are designed for a particle reservoir. In all these recovery processes, the interfacial tension (IFT) between the oil phase and the water phase is needed.

Another important factor is that, during water flooding, the water phase bypasses the oil in the reservoir (Figure 6.1). What this implies is that, if one injects water into the reservoir to push the oil, most of the water passes around the oil (the bypass phenomena) and comes up without being able to push the oil.

The pressure difference to push the oil drop may be larger than that to push the water, thus leading to the so-called bypass phenomena. In other words, as water flooding is performed, due to bypass, there is less oil produced, while more water is pumped back up with the oil.

The use of surfactants and other surface-active substances leads to the reduction of $\gamma_{oilwater}$ as described in Figure 6.1. The pressure difference at the oil blob entrapped in a reservoir and the surrounding aqueous phase will be

$$\Delta P_{oilwater} = 2 \, IFT \, (1/R_1 \; B \; 1/R_2) \tag{6.2}$$

Thus, by decreasing the value of interfacial tension (with the help of SAAs; from 50 mN/m to 1 mN/m), the pressure needed for oil recovery would be decreased.

In the water-flooding process, mixed emulsifiers are used. Soluble oils are used in various oil-well-treating processes, such as the treatment of water injection wells to improve water injectivity and to remove water blockage in producing wells. The same method is useful in different cleaning processes with oil wells. This is known to be effective since water-in-oil microemulsions are found in these mixtures, and with high viscosity. The micellar solution is composed essentially of hydrocarbon, aqueous phase, and surfactant sufficient to impart micellar solution characteristics to the emulsion. The hydrocarbon is crude oil or gasoline. Surfactants are alkyl aryl

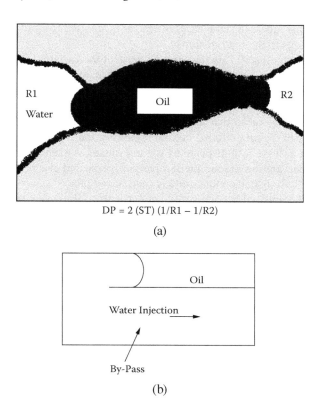

$$DP = 2 (ST) (1/R1 - 1/R2)$$

(a)

(b)

FIGURE 6.1 Water bypass in an oil reservoir: (a) oil would exhibit $\Delta P = 2\gamma(1/R_1 - 1/R_2)$; (b) water will bypass the oil trapped in the reservoir.

sulfonates, more commonly known as petroleum sulfonates. These emulsions may also contain ketones, esters, or alcohols as cosurfactants.

Another capillary phenomenon to consider is that the pores in the reservoirs are not perfectly circular. In the case of square-shaped pores, bypass is another possibility in the corners, but is not found in circular-shaped pores (Birdi et al., 1988).

6.2.1 Oil Spills and Cleanup Processes in Oceans

The treatment of oil spills is much dependent on surface chemistry principles. Oil on sea surfaces will be exposed to various parameters:

1. Loss by evaporation
2. Loss by sinking to the bottom (as such, or in conjunction with solids)
3. Emulsification

Oil spills are treated by various methods, depending on the region (in warmer seas or in a cold climate). The light fluids of oil will evaporate into the air. The oil that has adsorbed on solid suspension will sink to the bottom. The remaining oil is skimmed off by suitable machines. In some cases, surfactants are used to emulsify the oil, and this emulsion sinks to the bottom slowly.

However, no two oil spills are the same because of the variation in oil types, locations, and weather conditions involved. However, broadly speaking, there are four main methods of response.

1. Leave the oil alone so that it breaks down by natural means. If there is no possibility of the oil polluting coastal regions or marine industries, the best method is to leave it to disperse by natural means. A combination of wind, sun, current, and wave action will rapidly disperse and evaporate most oils. Light oils will disperse more quickly than heavy oils.
2. Contain the spill with booms and collect it from the water surface using skimmer equipment. Spilt oil floats on water and initially forms a slick that is a few millimeters thick. There are various types of booms that can be used either to surround and isolate a slick or to block the passage of a slick to vulnerable areas such as the intake of a desalination plant, fish farm pens, or other sensitive locations. Boom types vary from inflatable neoprene tubes to solid but buoyant material. Most rise up about a meter above the waterline. Some are designed to sit flush on tidal flats, while others are applicable to deeper water and have skirts that hang down about a meter below the waterline. Skimmers float across the top of the slick contained within the boom and suck or scoop the oil into storage tanks on nearby vessels or on the shore. However, booms and skimmers are less effective when deployed in high winds and high seas.
3. Use dispersants to break up the oil and speed its natural biodegradation. Dispersants act by reducing the surface tension that inhibits oil and water from mixing. Small droplets of oil are then formed, which helps promote rapid dilution of the oil by water movements. The formation of droplets also increases the oil surface area, thus increasing its exposure to natural evaporation and bacterial action. Dispersants are most effective when used within an hour or two of the initial spill. However, they are not appropriate for all oils and all locations. Successful dispersion of oil through the water column can affect marine organisms such as deep-water corals and sea grass. It can also cause oil to be temporarily accumulated by subtidal seafood. Decisions on whether or not to use dispersants to combat an oil spill must be made in each individual case. The decision will take into account the time since the spill, the weather conditions, the particular environment involved, and the type of oil that has been spilt.
4. Introduce biological agents to the spill to hasten biodegradation. Most of the components of oil washed up along a shoreline can be broken down by bacteria and other microorganisms into harmless substances such as fatty acids and carbon dioxide. This action is called biodegradation.

Oil spill technology is very advanced and can operate under very divergent conditions (from the cold seawaters to the tropic seas). The biggest difference arises from the oil, which may contain varying amounts of heavy components (such as tar).

6.3   APPLICATIONS IN CLEANING PROCESSES (SURFACE AND COLLOIDAL ASPECTS)

In almost all cleaning processes, unwanted materials (grease, dirt, color, bacteria) need to be removed from surfaces or cloth (cotton, synthetics, wool).

6.3.1   DETERGENCY

The detergent industry is a major field where surface and colloidal chemistry principles have been applied extensively. In fact, some detergent manufacturers have been involved in highly sophisticated research and development for many decades in developing numerous patented processes.

Dirt adheres to fabric by way of different forces (such as van der Waals and electrostatic). Some components of dirt are water soluble, and some are water insoluble. Commercial detergents are designed specifically for these particular processes to achieve various effects:

1. Water should be able to wet the fibers as completely as possible. This is achieved by lowering the surface tension of the washing water, which thus lowers the contact angle. The low value of surface tension also makes the washing liquid to be able to penetrate the pores (if present) since, from the Laplace equation, it can be seen that the pressure needed would be much lower. For example, if the pore size of fabric (any modern type such as microcotton and Gortex) is 0.3 µm, then it will require a certain pressure ($= \Delta P = 2\,\gamma/R$) in order for water to penetrate the fibers. In the case of water ($\gamma = 72$ mN/m) and using a contact angle of 105°, we obtain

$$\Delta P = 2\ (72\ 10^3)\ Cos\ (105)/0.3\ 10^6 = 1.4\ bar \qquad (6.3)$$

2. The detergent then interacts with the dirt–soil complex to start the process of the latter's removal from the fibers and dispersion into the washing water. In order to be able to inhibit the soil once removed from readsorbing into the clean fiber, polyphosphates or similar suitable inorganic salts should be used. These salts also increase the pH (around 10) of the washing water. In some cases, suitable polymeric antiredeposition substances (such as carboxymethyl cellulose) are preferred.
3. After the fabric is clean, special brighteners (fluorescent substances) can be applied that give a bluish haze to the fabric. This enhances the whiteness (by depressing the yellow tinge). Additionally, these also enhance the color perception. Brighteners as used for cotton are different from those used for synthetic fabrics.

Hence, the washing process is a series of well-designed steps that the industry has enhanced with much information and state-of-the-art technology. Washing technology has changed over the years as demands and preferences have changed. Washing machines are now designed to operate in conjunction with the soap industry. The mechanical movement and agitation of the machines is coordinated with the soap/detergent characteristics.

Typically, there are five different compositions of laundry detergents, shampoo, and dishwashing powder:

	Laundry	Shampoo	Dishwashing
Detergent (alkyl sulfate)	10–20	25	—
Soaps	5	—	—
Nonionics	5–10	1–5	
Inorganic salts (polyphosphate, silicates)	30–50	50	—
Optical brighteners	<1	—	—

It is worth noticing that the aim of detergents in these different formulations is different in each case. In other words, detergents are today tailor-made for each specific application (Ruiz, 2008). The detergents in shampoo should give stable foam in order to increase the cleaning effect. On the other hand, laundry detergents in dishwashing should exhibit a lower surface tension but almost no foaming (because foaming would reduce the cleaning effect). Hence, in dishwashing formulations, nonionics are used, which are not very soluble in water and thus produce very little (or no) foam. These are sometimes of type EOEOEOPOPOPO (ethylene oxide (EO)–propylene oxide (PO)). The propylene group behaves as *apolar,* and the oxide group behaves (through hydrogen bonding) as the *polar* part. These EOPO types can be tailormade by combining various ratios of EO:PO in the surfactant molecule. In some cases, even butylene-oxide groups have been used.

Additionally, in the case of shampoos, other criterion has to be strictly controlled. Shampoos are used to wash hair and are designed to remove dirt without damaging the hair. They should also have no skin or eye irritation effect. Baby shampoos are particularly manufactured using surfactants that exhibit minimum eye irritation.

Soil consists mainly of particulate, greasy matter. Detergents are supposed to keep the soil suspended in the solution and restrict redeposition. Tests also show that detergents stabilize suspensions of carbon or other solids such as manganese oxide in water. This suggests that detergents adsorb on the particles. Detergents add redeposition controllers such as carboxymethylcellulose.

In the early age of detergent usage during the 1960s, too much sewage treatment showed foaming problems. Later, detergents were used with better degradation properties and better control. For example, straight-chain alkyls were more degradable than branched alkyl chains.

6.3.2 WATER REPELLENCY OF MATERIALS

In many cases, such as with umbrellas and raincoats, the material used must be non-wetting in water. As mentioned earlier, if the contact angle is

Larger than 90°, a nonwetting solid surface is present.
Less than 90°, a wetting solid is present.

This is a very general statement but serves as a useful guideline for investigations. For water to penetrate fabrics, the magnitude of θ has been found to be close to zero. This can be achieved by using detergents. However, to achieve water repellency, the magnitude of θ has to be as large (i.e., >90°) as possible. If θ < 90°, then $Cos(\theta)$ is positive, and the liquid will penetrate a fabric. On the other hand, if θ >90°, the sign of $Cos(\theta)$ is negative, and the liquid will not penetrate the material. This should be compared to the capillary rise (or fall) of different liquids in a glass tubing (see Chapter 2).

In Young's equation, the quantity $(\gamma_S - \gamma_{SL})$ can be made negative. This is achieved by coating the solid material with some suitable material (such as Teflon). That means γ_c should be reduced to less than 30 mN/m.

6.4 EVAPORATION RATES OF LIQUID DROPS

In many natural (raindrops, fogs, river water fall) and industrial systems (sprays, oil combustion engines, cleaning processes), one encounters liquid drops. The rate of evaporation of liquid from such drops can be of importance in the function of these systems.

Extensive investigations on evaporation of liquid drops (free-hanging drops, drops placed on solid surfaces) have been reported in the current literature (Birdi, 2002). These drops have been analyzed as a function of

Liquid (water or organic)
Solids (plastics, glass)
Contact angle (θ)
Height and diameter and volume
Weight

In these analyses, some assumptions have been made as regards the shape of the drops. The most accurate data obtained is when using the weight method. Different analyses showed that the rate of evaporation was linearly dependent on the radius of the drop. Further, the contact angle of the water drop on Teflon (i.e., a nonwetting surface) remained constant under evaporation. On the other hand, the contact angle decreased as the water drop evaporated on glass (i.e., a wetting surface).

6.5 ADHESION (GLUES) (SOLID1–SOLID2)

There are many kinds of materials that need to be joined together by using glues or adhesives, both synthetic and organic, in manufacturing, the field of arts and crafts, and everyday life, such as

Plastic on glass
Metal to glass
Wood to wood (furniture, housing, boats)
Airplane and windmill wings

Energy is needed to break such a contact, more so with some joined items than others. If we consider the adhesion of plastic on glass, the highest adhesion will be obtained if the adhesive fills all the valleys and crevices of each adhered body surface. This will remove any air pockets that do not contribute to adhesion. The role of the adhesive or glue is to provide mechanical interlocking of the adhesive molecules.

The strength of the bond is dependent upon the quality of this interlocking interface. For achieving optimum bonding, chemical or physical abrading is used. The abrading process creates many useful properties at the solid surface: enhancement of mechanical interlocking, creation of a clean surface, formation of a chemically reactive surface, and an increase in surface rate (a smooth surface has lower surface area than a rough surface). Diffusion bonding is a form of mechanical interlocking that occurs at the molecular level in polymers.

The science behind bonding technology is very extensive. A brief description along with some examples is given in the following text. It is important to prime the surfaces of the layers to be bonded, that is, cover the surfaces with a dilute solution of the adhesive mixed with an organic solvent to obtain a dried film of thickness between 0.0015 and 0.005 mm.

Another example is concerned with epoxy adhesives. Epoxies are best, but an alloy such as epoxy–phenolic or epoxy–polysulfone may offer improved peeling resistance.

Dilute the adhesive until it has a lower surface tension than either of the bonding surfaces. The surface tensions can be compared by using a wetting test (i.e., by wetting the surface with the adhesive and measuring the contact angle). A low contact angle (<90) indicates good wetting and an appropriate adhesive.

Many theories have been developed to explain the process of bonding in adhesive structures. According to the mechanical bonding theory, an adhesive needs to fill the valleys and crevices of each adherend (body to be bonded) and displace trapped air to work effectively. Adhesion is the mechanical interlocking of the adhesive and the adherend together, and the overall strength of the bond is dependent upon the quality of this interlocking interface. This can be chemical or physical abrading for optimum bonding. Abrading the adherend does the following:

1. Enhances mechanical interlocking.
2. Creates a clean corrosion-free surface.
3. Forms a chemically reactive surface.
4. Increases the bonding surface area.

Diffusion bonding is a form of mechanical interlocking that occurs at the molecular level in polymers. The adsorption mechanism theory suggests that bonding is the process of intermolecular attraction (van der Waals bonding or permanent dipole, for example) between the adhesive and the adherend at the interface. An important factor in the strength of the bond, according to this theory, is the wetting of the adherend by the adhesive. Wetting is the process in which a liquid spreads onto a solid surface and is controlled by the surface energy of the liquid–solid interface versus the liquid–vapor and the solid–vapor interfaces. In a practical sense, to wet a solid surface, the adhesive should have a lower surface tension than the adherend.

In some systems with *charged surfaces,* the electrostatic forces will have to be considered. Electrostatic forces may also be a factor in the bonding of an adhesive to an adherent. These forces arise from the creation of an electrical double layer of separated charges at the interface and are believed to be a factor in the resistance to separation of the adhesive and the adherend. Adhesives and adherends that contain polar molecules or permanent dipoles are most likely to form electrostatic bonding, according to this theory, which has been developed to explain the curious behavior of the failure of bonded materials. Upon failure, many adhesive bonds break not at the adhesion interface but slightly within the adherend or the adhesive, adjacent to the interface. This suggests that a boundary layer of weak material is formed around the interface between the two media. In the following text, some mechanisms of adhesive failure are described.

The two predominant mechanisms of failure in adhesively bonded joints are adhesive failure or cohesive failure. Adhesive failure is the interfacial failure between the adhesive and one of the adherends. It indicates a weak boundary layer, often caused by improper surface preparation or adhesive choice. Cohesive failure is the internal failure of either the adhesive or, rarely, one of the adherends.

Ideally, the bond will fail within one of the adherends or the adhesive. This indicates that the maximum strength of the bonded materials is less than the strength of the adhesive strength between them. Usually, the failure of joints is neither completely cohesive nor completely adhesive. It is thus obvious that, for good bonding, the surfaces need to be clean. Any dirt, grease and lubricants, water or moisture, and weak surface scales need to be removed. Solvents are used to clean the soil from the solid surfaces of the adherend by using an organic solvent without affecting any physical property. The following are different procedures that have been found to be useful:

a. Vapor degreasing is a good procedure for surface cleaning.
b. Solvent wiping or immersion or spraying is a useful method.
c. Ultrasonic vapor degreasing is used as an effective method.

The most convenient method is ultrasonic treatment with a subsequent solvent rinse of the surface. Other intermediate procedures are abrasive scrubbing, filing, or detergent cleaning.

Cleaning can be a chemical treatment that includes acid or alkaline etching of the adherend surface. The etching process, especially, removes stubborn oxides (such as on aluminum or iron) and roughens the surface on a microscopic scale.

Priming is the process of applying a dilute solution of the adhesive mixed with an organic solvent on the adherend to a dried film thickness between 0.0015 and 0.005 mm (0.00006 to 0.002 in). Priming protects the surface from oxidation, improves wetting, helps prevent adhesive peeling, and serves as a barrier layer to prevent undesirable reactions between the adhesive and the adherent.

An adhesive is the basic substance that, when brought into contact with the hardener, after a chemical reaction, former becomes an effective adherent. The rate of adhesives is determined by appropriate additives. These include solvents that reduce the thickness of adhesives and also penetrate surfaces. In some cases, fillers are used to enhance adhesion and also to reduce costs.

Adhesives have been classified according to their applications. *Epoxy* is a compatible adhesive for stainless steel and ceramics, as used in the piezoelectric motor industry. Epoxy adhesives of different characteristics are found. Epoxy–polysulfone and epoxy–phenolics both exhibit thermosetting properties, providing stronger bonding properties with ceramics.

7 Colloidal Systems

7.1 INTRODUCTION

In this chapter, the very widespread industrial application of colloid chemistry will be described. Surprisingly enough, mankind has been aware of colloids for many thousands of years. The old Egyptian and Mayan civilizations, without cement, used their knowledge about adhesion (between blocks of stones) when building pyramids.

Our lives are a continuous encounter with solid particles of different sizes, ranging from stones on a beach, sand particles, or dust floating around in the air, without our being especially aware that there exists a special relation between particle size (surface area) and their characteristics. The rather small particles in the range of size from 50 Å to 50 μm are called *colloids*. The most obvious difference is apparent between sand particles and dust particles. It is fascinating to observe how dust or other fine particles remain in suspension in the air. In the 19th century, it was observed under the microscope (Brown) that small microscopic particle suspended in water made some erratic movements (as if hit by some other neighboring molecules). Since then this has been called *Brownian motion*. This erratic motion arises from the kinetic movement of the surrounding water molecules. Thus, colloidal particles would remain suspended in solution through Brownian motion only if the gravity forces did not drag these to the bottom (or top).

If sand is tossed into the air, the particles fall to the earth rather quickly. On the other hand, in the case of the talcum particles, these stay floating in the air for a long time. These differences will be described later. The size of particles may be considered from the following data:

Colloidal dispersions	10 nm–10 μ
Mist/fog	0.1 μ–10 μ
Pollen/bacteria	0.1 μ–10 μ
Oil in smoke/exhaust	1 μ–100 μ
Virus	10 nm–10 μ
Polymers/macromolecules	0.1 nm–100 nm
Micelles	0.1 nm–10 nm
Vesicles	1 μ–1000 μ

The stability of colloidal systems is subject to the state of their configuration, very roughly comparable to a bucket that is stable when standing up but if tilted beyond a certain angle, topples and comes to rest on its side (Figure 7.1).

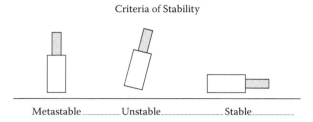

FIGURE 7.1   Stability criteria of any colloidal system: metastable–unstable–stable states.

A colloidal suspension may be unstable and exhibit separation of particles within a very short time, or it may be stable for a very long time, such as for over a year. In between, a *metastable* state can be found. This is an oversimplified example, but it shows that any colloidal system should be analyzed following these three criteria.

As an example, consider a wastewater treatment process. The wastewater with colloidal particles is a stable suspension. However, by treating it with pH control, electrolyte concentration, etc., the stability of the system can be altered as shown in Figure 7.1.

7.1.1 Van der Waals Forces

In colloidal systems, van der Waals forces play a prominent role. When any two particles (neutral or with charges) come very close to each other, the van der Waals forces will be strongly dependent on the surrounding medium. In a vacuum, two identical particles will always exhibit an attractive force. On the other hand, if two different particles are present in a medium (in water), then there may be repulsion forces. This can be due to one particle adsorbing with the medium more strongly than with the other particle. One example will be silica particles in water medium and plastics (as in wastewater treatment). It is critical to understand under what conditions it is possible that colloidal particles remain suspended. For example, if paint aggregates in the container, then it is obviously useless for its intended purpose.

When solid (inorganic) particles are dispersed in an aqueous medium, ions are released in the medium. The ions released from the surface of the solid are of opposite charge. This can be easily shown when glass powder is mixed in water, and conductivity is seen to increase with time. The presence of the same charge on particles in close proximity gives repulsion, which keeps the particles apart (Figure 7.2).

The positive–positive particles will show repulsion. On the other hand, the positive–negative particles will attract each other. The ion distribution will also depend on the concentration of any counterions or coions in the solution. Even glass, when dipped in water, exchanges ions with its surroundings. Such phenomena can be easily investigated by measuring the change in the conductivity of the water.

The force F_{12}, acting between these oppositely charges, is given by Coulomb's law, with charges q_1 and q_2 separated at a distance R_{12} in a dielectric medium D_e:

$$F_{12} = (q_1 \, q_2)/(4 \, \pi \, D_e \, \varepsilon_o \, R_{122}) \tag{7.1}$$

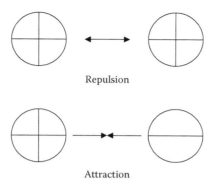

Repulsion

Attraction

FIGURE 7.2 Solid with charges (positive–positive [repulsion] or positive–negative [attraction]).

where the force would be attractive between oppositely charges, while repulsive in the case of similar charges. Because the D_e of water is very high (80 units) as compared to the D_e of air (ca. 2), we will expect very high dissociation in water, while there is hardly any dissociation in air or organic liquids. Let us consider the F_{12} for Na^+ and Cl^- ions (with charge of $1.6 \times 10^{-19} C = 4.8 \times 10^{-10}$ esu) in water ($D_e = 74.2$ at $37°C$), and at a separation (R_{12}) of 1 nm:

$$F_{12} = (1.6 \ 10^{-19})(1.6 \ 10^{-19})/[(4 \ \Pi \ 8.854 \ 10^{-12}) \ (10^{-9})(74.2)] \qquad (7.2)$$

where ε_o is 8.854×10^{-12} $kg^{-1}m^{-3}$ s^4 A^2 (J^{-1} C^2 m^{-1}). This gives a value of F_{12} of -3.1×10^{-21} J/molecule or -1.87 kJ/mol.

Another very important physical parameter one must consider is the size distribution of the colloids. A system consisting of particles of the same size is called a *monodisperse*. A system with different sizes is called *polydisperse*. It is also obvious that systems with monodisperse will exhibit different properties from those of polydispersed systems. In many industrial application (such as coating on tapes used for recording music and coatings on CDs or DVDs), latter kind of quality of coatings is needed.

The methods used to prepare monodisperse colloids aim to achieve a large number of critical nuclei in a short interval of time. This induces all equally sized nuclei to grow simultaneously, thus producing a monodisperse colloidal product.

7.1.2 Colloids Stability (DLVO Theory)

We need to understand under which conditions a colloidal system will remain dispersed (and under which it will become unstable). Knowing how colloidal particles interact with one another makes possible an appreciation of the experimental results for phase transitions in such systems as found in various industrial processes. It is also necessary to know under which conditions a given dispersion will become unstable (coagulation). For example, one needs to apply coagulation in wastewater treatment so that most of the solid particles in suspension can be removed. Any two particles coming close to each other, will produce different forces.

7.2   ATTRACTIVE FORCES–REPULSIVE FORCES

If the attractive forces are larger than the repulsive forces, then the two particles will merge together. However, if the repulsion forces are larger than the attractive forces, then the particles will remain separated. The medium in which these particles are present is also a factor. pH and ionic strength (i.e., concentration of ions) are especially found to exhibit very specific effects.

Different forces of interest can be

- Van der Waals
- Electrostatic
- Steric
- Hydration
- Polymer–polymer interactions (if polymers are involved in the system)

In many systems, one may add large molecules (polymers), which, when adsorbed on the solid particles, will impart a special kind of stability criteria.

It is well known that neutral molecules such as alkanes attract one another, mainly through van der Waals forces. Van der Waals forces arise from the rapidly fluctuating dipoles moment (10^{15} S^{-1}) of a neutral atom, which leads to polarization and consequently to attraction. This is also called the *London potential* between two atoms in a vacuum, and is given as

$$V_{vdw} = -(L_{11}/\mathbf{R}^6) \tag{7.3}$$

where L_{11} is a constant that depends on the polarizability and the energy related to the dispersion frequency, and $\mathbf{R}$ is the distance between the two atoms. Since the London interactions with other atoms may be negated as an approximation, the total interaction for any macroscopic bodies may be estimated by a simple integration.

When two similarly charged colloid particles, under the influence of the EDL, come close to each other, they will begin to interact. The potentials will detect one another, and this will lead to various consequences. The charged molecules or particles will be under both van der Waals and electrostatic interaction forces. The van der Waals forces, which operate at a short distance between particles, will give rise to strong attraction forces. The potential of the mean force between colloid particle in an electrolyte solution plays a central role in the phase behavior and the kinetics of agglomeration in colloidal dispersions. This kind of investigation is important in these various industries:

- Inorganic materials (ceramics, cements)
- Foods (milk)
- Biomacromolecular systems (proteins and DNA)

The DLVO (*Derjaguin–Landau–Verwey–Overbeek*) theory describes the stability of a colloidal suspension as mainly dependent on the distance between the particles

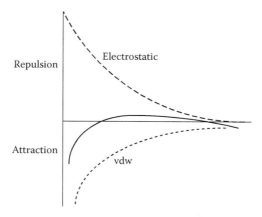

FIGURE 7.3 Variation of repulsion and attraction forces versus distance between two particles (schematic).

(Adamson and Gast, 1997). DLVO theory has been modified in later years, and different versions are found in the current literature.

The electrostatic forces will give rise to repulsion at large distances (Figure 7.3). This arises from the fact that the electrical charge–charge interactions take place at a large distance of separation. The resultant curve is shown (schematic) in Figure 7.3. The barrier height determines the stability with respect to the quantity $k\,T$, the kinetic energy. DLVO theory predicts, in most simple terms, that if the repulsion potential (Figure 7.4) exceeds the attraction potential by a value

$$W \gg k\,T \tag{7.4}$$

then the suspension will be stable. On the other hand, if

$$W <= k\,T \tag{7.5}$$

then the suspension will be unstable, and it will coagulate. It must be stressed that DLVO theory does not provide a comprehensive analysis. It is basically a very useful tool for analysis of complicated systems. Especially, it is a useful guidance theory in any new application or industrial development.

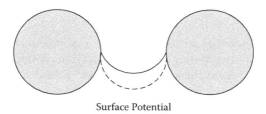

Surface Potential

FIGURE 7.4 Variation of EDL between two charged particles with different ion concentrations (low; high).

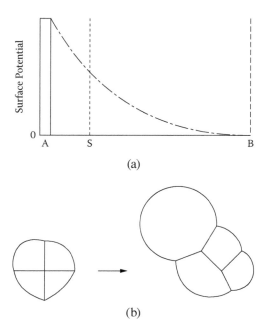

(a)

(b)

FIGURE 7.5 Variation of (a) surface potential, ψ_0, and (b) ions (sodium/chloride) versus distance from the solid.

7.2.1 Charged Colloids (Electrical Charge Distribution at Interfaces)

The interactions between two charged bodies will be dependent on various parameters (e.g., surface charge, electrolyte in the medium, charge distribution; see Figure 7.4). The distribution of ions in an aqueous medium needs to be investigated in such charged colloidal systems. This observation indicates that the presence of charges on surfaces means that there exists a potential that needs to be investigated. On the other hand, in the case of neutral surfaces, one has only the van der Waals forces to be considered. This was clearly seen in the case of micelles, where the addition of small amounts of NaCl to the solution showed

* A large decrease in CMC in the case of ionic surfactant
* Almost no effect in nonionic micelles (since in these micelles there are no charges or EDL)

Electrostatic and EDL forces are found to play a very important role in a variety of systems in science and engineering. It would be useful to consider a specific example in order to understand these phenomena. Let us take a surface with positive charge that is suspended in a solution containing positive and negative ions. There will be a definite surface potential, ψ_0, which decreases to a value zero as one moves away into solution (Figure 7.5).

It is obvious that the concentration of positive ions will decrease as one approaches the surface of the positively charged surface (charge–charge repulsion). On the other hand, the oppositely charged ions, negative, will be strongly attracted toward the surface. This produces the so-called Boltzmann distribution:

$$n^- = n_0 \, e^{(z \, \varepsilon e \, \psi/k \, T)} \tag{7.6}$$

$$n^+ = n_0 \, e^{-(z \, \varepsilon e \, \psi/k \, T)} \tag{7.7}$$

This shows that positive ions are repelled, while negative ions are attracted to the positively charged surface. At a reasonably far distance from the particle, $n^+ = n^-$ (as required by the electroneutrality). Through some simple assumptions, one can obtain an expression for $\psi \, (r)$, as a function of distance, r, from the surface as

$$\psi \, (r) = z \, e/(\mathbf{D} \, r) \, \varepsilon e - \kappa \, r \tag{7.8}$$

where κ is related to the ion atmosphere around any ion. In any aqueous solution, when an electrolyte, such as NaCl, is present, it dissociates into positive (Na^+) and negative (Cl^-) ions. Because of requirement of electroneutrality (i.e., there must be same positive and negative ions), each ion is surrounded by an appositively charged ion at some distance. Obviously, this distance will decrease with increasing concentration of the added electrolyte. The expression $1/\kappa$ is called the Debye length.

As expected, the D-H theory tells us that ions tend to cluster around the central ion. A fundamental property of the counterion distribution is the thickness of the ion atmosphere. This thickness is determined by the quantity Debye length or Debye radius ($1/\kappa$). The magnitude of $1/\kappa$ has dimension in centimeters, as follows:

$$\kappa = ((8 \, N \, 2)/(1000 \, \mathbf{k}_B \, T)\tfrac{1}{2} I^{1/2} \tag{7.9}$$

The values of $\mathbf{k}_B = 1.38 \, 10^{-23}$ J/molecule K, $e = 4.8 \, 10^{-10}$ esu. Thus, the quantity $\mathbf{k}_B T/e$ = 25.7 mV at 25°C. As an example, with a 1:1 ion (such as NaCl, KBr) with concentration 0.001 M, one gets the value of $1/\kappa$ at 25°C (298 K):

$$1/\kappa = (78.3 \, 1.38 \, 10^{-16} \, 298)/(2 \, 4 \, \Pi \, 6.023 \, 10^{17})(4.8 \, 10^{-10})^2)^{0.5}$$

$$= 9.7 \, 10^{-7} \, cm$$

$$= 97 \, \text{Å} \tag{7.10}$$

The expression, in equation, can be rewritten as

$$\psi \, (r) = \psi_0 \, (r) \, \exp(-\kappa \, r) \tag{7.11}$$

which shows the change in $\psi \, (r)$ with the distance between particles (r). At a distance $1/\kappa$, the potential has dropped to ψ_0. This is accepted as corresponding with the thickness of the double layer. This is an important analysis since the particle–particle interaction is dependent on the change in $\psi \, (r)$. The decrease in $\psi \, (r)$ at the Debye length is different for different ionic strength (Figure 7.6).

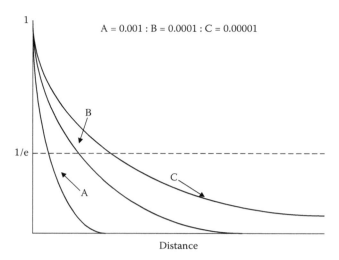

FIGURE 7.6 Variation (decrease) in electrostatic potential with distance of separation as a function of electrolyte concentration (ionic strength).

TABLE 7.1
Debye Length ([1/k] nm) in Aqueous Solutions

Salt Concentration Molal	1:1	1:2	2:2
0.0001	30.4	17.6	15.2
0.001	9.6	5.55	4.81
0.01	3.04	1.76	1.52
0.1	0.96	0.55	0.48

The data in Table 7.1 show values of D-H radius in various salt concentrations. The magnitude of $1/\kappa$ decreases with I and with the number of charges on the added salt. This means that the thickness of the ion atmosphere around a reference ion will be much compressed with increasing value of I and z_{ion}.

A trivalent ion such as Al^{3+} will compress the double layer to a greater extent in comparison with a monovalent ion such as Na^+. Further, inorganic ions can interact with the charge surface in one of two distinct ways:

1. Nonspecific ion adsorption, where these ions have no effect on the isoelectric point
2. Specific ion adsorption, which gives rise to change in the value of the isoelectric-point

Under those conditions, where the magnitude of $1/\kappa$ is very small (e.g., in high electrolyte solution), one can write:

$$\psi = \psi_o \exp - (\kappa\, x) \qquad (7.12)$$

where x is the distance from the charged colloid. The value of ψ_o is found to be 100 mV (in the case of monovalent ions) (= 4 **kB** T/z e). Experimental data and theory show that the variation of ψ is dependent on the concentration and the charge of the ions (Figure 7.7).

These data show that the surface potential drops to zero at a faster rate if the ion concentration (C) increases, and that the surface potential drops faster if the value of z goes from 1 to 2.

7.2.2 ELECTROKINETIC PROCESSES

In the following text, let us consider what happens if the charged particle or surface is under dynamic motion of some kind. Further, there are different systems under which electrokinetic phenomena are investigated. These systems are

1. *Electrophoresis*: This system refers to the movement of the colloidal particle under an applied electric field.
2. *Electroosmosis*: This system is one where a fluid passes next to a charged material. This is actually the complement of electrophoresis. The pressure needed to make the fluid flow is called the *electroosmotic pressure*.
3. *Streaming potent*: If fluid is made to flow past a charged surface, then an electric field is created, which is called *streaming potential*. This system is thus the opposite of electroosmosis.
4. *Sedimentation potential*: A potential is created when charged particles settle out of a suspension. This gives rise to sedimentation potential,

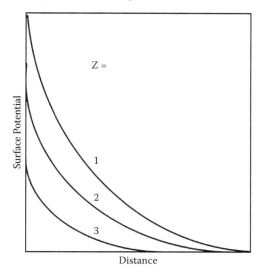

FIGURE 7.7 Variation of surface potential (schematic) in the diffuse double layer as a function of charge on the ions (z).

which is the opposite of streaming potential. The reason for investigating electrokinetic properties of a system is to determine a quantity known as the *zeta potential*.

Electrophoresis is the movement of an electrically charged substance under the influence of an electric field. This movement may be related to fundamental electrical properties of the body under study and the ambient electrical conditions by the following equation. F is the force, q is the charge carried by the body, and E is the electric field:

$$F_e = q\,E \qquad (7.13)$$

The resulting electrophoretic migration is countered by forces of friction such that the rate of migration is constant in a constant and homogeneous electric field:

$$Ff = \mathbf{v}\,f_r \qquad (7.14)$$

where **v** is the velocity, and fr is the frictional coefficient.

$$Q\,E = v\,fr \qquad (7.15)$$

The electrophoretic mobility µ is defined as

$$\mu = \mathbf{v/E} = \mathbf{q/fr} \qquad (7.16)$$

The foregoing expression applies only to ions at a concentration approaching 0 and in a nonconductive solvent. Polyionic molecules are surrounded by a cloud of counterions that alter the effective electric field applied on the ions to be separated. This renders the previous expression a poor approximation of what really happens in an electrophoretic apparatus.

The mobility depends on both the particle properties (e.g., surface charge density and size) and solution properties (e.g., ionic strength, electric permittivity, and pH). For high ionic strengths, an approximate expression for the *electrophoretic mobility*, μ_e, is given by the Smoluchowski equation:

$$\mu_e = \varepsilon\,\varepsilon o\,\eta/n \qquad (7.17)$$

where ε is the dielectric constant of the liquid, εo is the permittivity of free space, η is the viscosity of the liquid, and η is the zeta potential (i.e., surface potential) of the particle.

7.2.3 Stability of Lyophobic Suspensions

Particles in all kinds of suspensions or dispersions interact with two different kinds of forces (e.g., attractive forces and repulsive forces). One observes that lyophobic suspensions (sols) must exhibit a maximum in repulsion energy in order to have a stable system. The total interaction energy, V (h), is given as (Scheludko, 1966; Bockris et al., 1980; Adamson and Gast, 1997; Chattoraj and Birdi, 1984; Birdi, 2002)

$$V(h) = V_{el} + V_{vdw} \tag{7.18}$$

where V_{el} and V_{vdw} are electrostatic repulsion and van der Waals attraction components. Dependence of the interaction energy $V(h)$ on the distance **h** between particles has been ascribed to coagulation rates as follows:

  a. During slow coagulation
  b. When fast coagulation sets in

The dependence of energy on **h** and V(h),

$$V(h) = [(64\ C\ RT\ \psi^2)/k\ \exp(-k\ \mathbf{h}) - H/(2\ h^2) \tag{7.19}$$

satisfies the requirements of this coagulation rate. For a certain ratio of constants, it has the shape shown in Figure. 7.8. For large values of **h**, V(h) is negative (attraction), following the energy of attraction V_{vdw}, which decreases more slowly with increasing distance ($\sim 1/h^2$). At short distances (small h), the positive component V_{el} (repulsion), which increases exponentially with decreasing **h**, $(\exp(-k\ h)$, can overcompensate V_{vdw} and reverse the sign of both dV(h)/dh and V(h) in the direction of repulsion. On further reduction of the gap (very small h), V_{vdw} should again predominate since

$$V_{el} = 64\ C\ RT\ \psi^2/k,\ \text{as h} \geq 0 \tag{7.20}$$

whereas the magnitude of V_{vdw} increases indefinitely when h > 0. There is thus a repulsion maximum in the function V(h), which can be easily found from the condition dV(h)/dh = 0. The transition from stability (a) to instability (c) as the electrolyte concentration, C, is increased as shown in Figure 7.8. Curve (b) corresponds to the onset of rapid coagulation. The choice of solution (maximum or minimum) does not present any difficulty since V(h) is positive for the maximum. The solution of

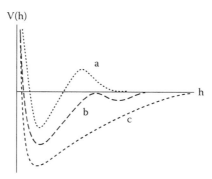

FIGURE 7.8 Variation of V(h) vs. h: (a) stable; (b) rapid coagulation; (c) unstable. (See text for details.)

Equation 7.20 with respect to V_{max} is involved, and will be not discussed here. It is, however, readily appreciated that, when the electrolyte concentration is increased, the magnitude of k in the exponent of V_{el} also increases (compression of diffuse layers), so that the maximum caused by it becomes lower. At a certain value of C, the curve V(h) will become similar to curve b in Figure 7.8 with $V_{max} = 0$. In accordance with all that has been said before, coagulation will become fast starting from this concentration. This is therefore the *critical concentration*, C_{cc}. In other words, the critical concentration (C_{cc}) can be estimated from simultaneous solution of following:

$$dV(h)/dh = 0 \text{ and } V(h) = 0 \qquad (7.21)$$

One can write the following:

$$dV(h)/dh = [-(64 \ C_{cc} \ RT \ \psi^2)/k \ \exp(-k_{cr} \ h_{cr}) + K/(h_{cr}^3) = 0 \qquad (7.22)$$

and

$$V(h) = [(64 \ C_{cc} \ RT \ \psi 2)/k_{cr} \ exp(-k_{cr} \ h_{cr}) - K/(2 \ h_{cr} \ 2) = 0 \qquad (7.23)$$

After expanding these expressions, as related to h and C, this becomes (Schulze–Hardy Rule for suspensions in water)

$$C_{cc} = 8.7 \ 10^{-39}/Z^6 \ A^2$$

$$C_{cc} \ Z^6 = \text{constant} \qquad (7.24)$$

where the constant includes (Hamaker constant is approximately $4.2 \ 10^{-19}$ J). Concentrations of ion to the sixth power of various valencies are inversely proportion to valency.

$$Z = 1:(2^6)0.016:(3^6)0.0014:(4^6)0.000244 \qquad (7.25)$$

The flocculation concentrations of mono-, di-, and trivalent gegen ions should, from this theory, be expected as

$$1: (1/2)^6 : (1/3)^6 \qquad (7.26)$$

It thus becomes obvious that the colloidal stability of charged particles is dependent on

  i. Concentration of electrolyte
 ii. Charge on the ions
iii. Size and shape of colloids
 iv. Viscosity

The critical concentration (critical coagulation concentration) is thus found to depend on the type of electrolyte used, as well as on the valency of the counterion. It is seen that divalent ions are 60 times as effective as monovalent ions, and trivalent ions are several hundred times more effective than monovalent ions. However, ions that specifically adsorb (such as surfactants) will exhibit different behavior.

Based on these observations, in the composition of washing powders, multivalent phosphates are used, for instance, to keep the charged dirt particles from attaching to the fabrics after having been removed. Another example is wastewater treatment where, for coagulation purposes, multivalent ions are used.

Streaming potential: The interface of a mineral (rock) in contact with an aqueous phase exhibits surface charge. The currently accepted model of this interface is the EDL model of Stern. Chemical reactions take place between the minerals and the electrolytes in the aqueous phase, which results in a net charge on the mineral. Water and electrolytes bound to the rock surface constitute the Stern (or Helmholtz) layer. In this region, the ions are tightly bound to the mineral, while away from this layer (the so-called diffuse layer), the ions are free to move about.

Since the distribution of ions (positive and negative) is even in the diffuse region, there is no net charge. On the other hand, in the Stern layer, there will be asymmetric charge distribution, and thus one will measure from zeta-potential data that the mineral exhibits a net charge.

7.3 KINETICS OF COAGULATION OF COLLOIDS

Colloidal solutions are characterized by the degree of stability or instability. This is related to the fact that both kinds of properties in everyday phenomena need to be understood. The kinetics of coagulation is studied using different methods. The number of particles, N_p, at a given time is dependent on the diffusion-controlled process. The rate is given by

$$-d\,N_p/dt = 8\,\pi\,D\,R\,N_p^2 \qquad (7.27)$$

where D is the diffusion coefficient, and R is the radius of the particle. The rate can rewritten as

$$-d\,N_p/dt = 4\,k_B T/3\,v\,N_p^2 = k_o\,N_p^2 \qquad (7.28)$$

where $D = k_B\,T/6\,\pi\,v\,R$, the Einstein equation is applied, and k_o is the diffusion-controlled constant.

In real systems, both stable colloidal systems (as in paints, creams) and unstable systems (as in wastewater treatment) are of interest. It is thus seen that, from DLVO considerations, the degree of colloidal stability will be dependent on the following factors:

1. Size of particles; larger particles will be less stable.
2. Magnitude of surface potential.
3. Hamaker constant (**H**).
4. Ionic strength.
5. Temperature.

The attraction force between two particles is proportional to the distance of separation and a Hamaker constant (specific to the system). The magnitude of $\mathbf{H}$ is of the order of 10^{-12} erg (Adamson and Gast, 1997; Birdi, 2002).

The DLVO theory is thus found to be useful to predict and estimate colloidal stability behavior. Of course, in such systems with many variables, a simplified theory is to be expected to fit all kinds of systems.

In the past decade, much development has taken place in regard to measuring the forces involved in these colloidal systems. In one method, the procedure used is to measure the force present between two solid surfaces at very low distances (less than micrometer). The system can operate under water, and thus the effect of addictives has been investigated. These data have provided verification of many aspects of the DLVO theory. Recently, the atomic force microscope (AFM) has been used to measure these colloidal forces directly (Birdi, 2002). Two particles are brought closer, and the force (nanoNewton) is measured. In fact, commercially available apparatus are designed to perform such analyses. The measurements can be carried out in fluids and under various experimental conditions (such as added electrolytes, pH, etc.).

7.3.1 FLOCCULATION AND COAGULATION OF COLLOIDAL SUSPENSION

It is known from common experience that a colloidal dispersion with smaller particles is more stable than one with larger particles. The phenomenon of smaller particles forming aggregates with larger particles is called *flocculation* or *coagulation*. For example, to remove insoluble and colloidal metal that is precipitated, flocculation is used. This is generally achieved by reducing the surface charges that give rise to weaker charge–charge repulsion forces. As soon as the attraction forces (vdW) become larger than the electrostatic forces, coagulation takes place. It is initiated by particle charge neutralization (by changing pH or other methods [such as charged polyelectrolytes]), which leads to aggregation of particles to form the larger size. This approach is based on to change an initially charged particle to a neutral particle:

Initial state: Charge–charge repulsion
Final state: Neutral–neutral (attraction)

Coagulation can also be brought about by adding suitable substances (coagulants) that are particular to a given system. This method reduces the effective radius of the colloid particle and leads to coagulation. Flocculation is the secondary process after coagulation, and this leads to very large particle (floccs) formation. Experiments show that coagulation takes place when the zeta potential is around ± 0.5 mV. Coagulants such as iron and aluminum inorganic salts are effective in most cases. In wastewater treatment plants, the zeta potential is used to determine coagulation and flocculation phenomena. Most of the solid material in wastewater is negatively charged.

7.4 DISPERSIONS OF SOLID PARTICLES IN FLUIDS

As described earlier, when a solid particle or a liquid drop is broken down in size the free energy of the system increases (because the magnitude of surface area per gram of solid increases). This happens because molecules in the bulk phase are brought to the surface, which needs energy (work). The change in the surface free energy is the product of the surface area produced (surface area increase) and the surface tension of the interface created. By changing the surface tension, one may produce different-sized particles for a given amount of energy input (such as grinding, rolling, or shaking). In the grinding process, an input of mechanical energy is needed to break down the particles. If surface tension is decreased (by adding a suitable surface-active agent), then, for the same input of mechanical energy, the size of particles will decrease with the decrease in surface tension. Moreover, if the grinding is performed under dry conditions, there will be a strong tendency for the particles formed to adhere to each other and create the well-known caking problems. As a related example, the size of crystals decreases when a saturated salt solution is cooled and a surfactant is added (thus reducing the crystal–solution interfacial surface tension).

The surface tension of the system can also be changed by grinding with liquid, thus decreasing interfacial tension. This gives rise to a variety of parameters since, by adding suitable chemicals (electrolytes or surface-active agents), one can modify the end-product properties. Conversely, the size of crystals formed from a supersaturated solution of a substance is related to the surface tension (at the solid–liquid interface). Thus, to obtain fine crystals, a suitable detergent is added, and thus, finer crystals are obtained.

A typical example of this is the production of glass fibers that are used for isolation. In order to keep the negatively charged glass fibers from strong adhesion, a cationic charged surface is sprayed with an active agent, which enhances the isolation by keeping the fibers from compact structure formation.

In the past few decades, a specific kind of colloidal system based on monodisperse size has been developed for various industrial applications. A variety of metal oxides and hydroxides and polymer lattices have been produced. Monodisperse systems are obviously preferred since their properties can be easily predicted. On the other hand, *polydisperse* systems will exhibit varying characteristics, depending on the degree of polydispersity.

7.5 APPLICATIONS OF COLLOID SYSTEMS

The stability of colloid suspensions is an important criteria in the manufacture of a large number of industrial products where these are the basic building blocks (food colloids, pollution control, emulsions, wastewater treatment).

An example of a food colloid is mayonnaise (a mixture of vegetable oil plus egg yolk and vinegar which is an emulsion of oil in water).

The electrostatic forces in many systems play a dominant role, such as the separation process (filtration) in wastewater treatment.

7.5.1 Wastewater Treatment and Control (Zeta Potential)

Wastewater contains different kinds of pollutants (dissolved substances, suspended particles), and is treated in suitable plants before the processed water is released.

Wastewater contains both soluble and insoluble substances that need to be removed. These substances that (solutes) are in either molecular (such as benzene, etc.) or ionic (such as Na^+, Cl^-, Mg^{++}, K^+, Fe^{++}, etc.) form. The concentrations are given in various units:

Weight/volume	$mg\ L^{-1}$; $kg\ m^{-3}$
Weight/w	$mg\ kg^{-1}$; parts per million (ppm); parts per billion (ppb)
Molarity	$mol\ L^{-1}$
Normality	equivalents L^{-1}

The specific unit used depends on the amounts present. The unit used for a trace amount of a solute, such as benzene, is given in ppm or ppb. The hardness of drinking water (mostly Na, Ca-Mg) concentration is given as $mg\ L^{-1}$. The typical values as found are in the range of less than $10\ mg\ L^{-1}$ (soft water) or over $20\ mg\ L^{-1}$ (hard water).

The presence of a net charge at the particle surface produces an asymmetric distribution of ions in the surrounding region. This means that the concentration of counterions close to the surface are higher than the ions with the same charge as the particle. Thus, an electrical double layer is measured around such a particle placed in water.

The solids can be removed by filtration and precipitation methods. The precipitation (of charged particles) is controlled by making the particles flocculate by controlling the pH and ionic strength. The latter gives rise to a decrease in charge–charge repulsion, and can lead to precipitation and removal of finely divided suspended solids. The most important factor that effects *zeta potential* is found to be pH. Therefore, all zeta-potential data must include a note on its pH. Imagine a particle in suspension with a negative zeta-potential. If more alkali is added to this suspension, then the particle will exhibit an increase in negative charge. On the other hand, if acid is added to the colloidal suspension, then the particle will acquire increasing positive charge. During this process, the particle will undergo a change from negative charge to *zero* charge (where the number of positive charge is equal to the negative charge (*point-of-zero-charge* [PZC]). In other words, one can control the magnitude and sign of the surface charge by a potential-determining ion.

The stability is dependent on the magnitude of electrostatic potential, ψ, at the surface of the colloid, ψ_0. The magnitude of ψ_0 is estimated by using the microelectrophoresis method. When an electric field is applied across an electrolyte, charged particles suspended in the electrolyte are attracted toward the electrode of opposite charge. Viscous forces acting on the particles tend to oppose this movement. When equilibrium is reached between these two opposing forces, the particles move with constant velocity. In this technique, the movement (or rather the speed) of a particle is observed under a microscope subjected to an electric field. The field is related to the applied voltage, **V**, divided by the distance between the electrodes (in centimeters). The velocity is dependent on the strength of the electric field or voltage gradient, the dielectric constant of the medium, viscosity, and zeta-potential. Commercially available electrophoresis instruments are used where the quartz cells designed for any specific system are available.

The magnitude of the zeta potential, ζ, is obtained from the following relation:

$$\zeta = \mu \, \eta / \varepsilon_o \, \mathbf{D} \tag{7.29}$$

where η is the viscosity of the solution, ε_o is the permittivity of the free space, and $\mathbf{D}$ is the dielectric constant. The velocity of a particle in a unit electric field is related to its electric mobility.

In another application, the magnitude of the zeta potential is measured as a function of added counterions. The variation in zeta potential is found to be related to the stability of the colloidal suspension. The results of a gold colloidal suspension (gold solute) are reported as follows:

Counterion	Velocity	Stability (Flocculation Character)
0 Al + 3	3(−)	Very high stability
20 10^{-6} mol	2(−)	Flocculates (4 h)
30 10^{-6} mol	0(zero)	Flocculates fast
40 10^{-6} mol	0.2(+)	Flocculates (4 h)
70 10^{-6} mol	1(+)	Flocculates slowly

These data show that the charge on the colloidal particles goes from negative to zero (when the particles do not show any movement) to positive at high counterion concentration. This is a very general picture. Therefore, in wastewater treatment plants, counterions are added until the movement is almost zero, and fast flocculation of pollutant particles can be achieved. The variation of ζ of silica particles has been investigated as a function of pH. The dissociation of the surface groups $-Si(OH)$ is also involved. In this process, the zeta potential is constantly monitored by using a suitable instrument. In some plants, this is carried out under continuous measurement.

In some biological systems with the charge–charge interactions between larger protein molecules, such as hemoglobin (mol. wt. 68,000), the aggregation becomes critical if surface charges change.

7.5.2 STERIC STABILIZATION OF SOLID OR LIQUID COLLOIDS

The stability of solid particles or liquid drops can be also controlled by using large molecules (*polymers*). The addition of a polymer will result in adsorption in a solid or penetration in a liquid (Figure 7.9).

The mechanism of polymer stabilization is many-faceted. It introduces following new parameters:

The colloid is imparted a different charge, depending on the polymer. In fact, if the polymer is neutral, then the colloid may even become neutral.
The size of the polymer imparts special stability.
The polymer affects the viscosity of the media.

The main advantages in using polymers for colloidal stability are

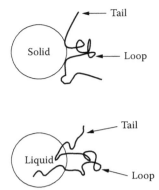

FIGURE 7.9 State of a polymer at (a) solid or (b) liquid drop. The polymer may be in three typical states: tail, loop, or coil (helix-coil).

When two colloids approach each other, the adsorbed polymers increase repulsion.

If the polymers interact, the two colloids will exhibit increased repulsion as the solvent molecules would push toward this area.

The polymers, after adsorption, change in conformation, and thus may increase stability.

However, one needs to find the suitable polymer for a given colloidal system. A typical example as found in biology is the stabilization of milk.

7.5.2.1 Other Applications of the Zeta Potential

In various kinds of industrial production, materials need to be treated with charged colloidal particles. In such systems, the value of the zeta-potential analyses are needed to control production. For example, in paper, adhesive, and synthetic plastics, colloidal clay can be used as filler. In oil drilling, clay colloidal suspensions are used. The zeta potential is controlled so as to avoid clogging the pumping process in the oil well. It has been found that, for instance, the viscosity of a clay suspension shows a minimum when the zeta potential is changed (with the help of pH from 1 to 7) from 15 to 35 mV. Similar observations have been reported in coal slurry viscosity. The viscosity was controlled by the zeta potential.

7.5.3 INDUSTRIAL APPLICATIONS

Colloids are found to play an important role in many aspects of industrial products. Therefore, there is much research devoted to the following aspects of colloids:

Production of well-defined colloidal products
Stability criteria of colloidal products
Disability criteria
New applications of novel colloidal preparations

Some industrial products in which the foregoing characteristics are of primary importance include paints, inks, food products, and pharmaceuticals.

The potential differences, ψ, at different phase boundaries, as mentioned before, have been found to have many industrial applications. The application of electrophoresis to the separation and purification of proteins has also been discussed. Both electrophoresis and electroosmosis have attained a certain amount of industrial application.

They are often used in the purification of natural colloids. For example, the colloids may be separated from uncharged or oppositely charged impurities by electrophoresis deposited on an electrode. The electrophoretic purification of clay and kaolin may serve as an example. The clay is made into a negatively charged sludge whose charge may be increased by the addition of small quantities of a base (potential-determining OH ions) so that the suspension stable colloid while the impurities settle out. The suspension is then subjected to electrophone; thus, the colloidal particles are deposited on a rotating metallic cylinder that serves as an anode in which they can be scraped off continuously. The clay is dehydrated during production.

The colloidal particles are often deposited on metallic electrodes in the form of adsorbed coatings. Rubber and graphite coatings can be formed in this way, using solvent mixtures (water–acetone) as the dispersion media. The advantage of this method is that additives can firmly be codeposited with, for example, rubber latex. Thermionic emitters for radio valves are produced in a similar manner. The colloidal suspensions of alkaline earth carbonates are deposited electrophoretically on the electrode and are later converted to oxides by using an ignition process.

8 Thin Liquid Films

8.1 INTRODUCTION

The formation and structure of *thin liquid films* (TLF; such as in foams or bubbles) is a fascinating phenomena that humans have studied over many decades. This structure is the closest one comes to observing molecules with the eye. TLF is thus the thinnest object one can see without the aid of any kind of microscope. One of the most commonly known thin liquid film structures is the soap bubble or bubbles formed on detergent solutions (as in dishwashing solutions). Everyone has enjoyed this formation and its display of rainbow colors. It may look as if a bubble formation and its characteristic stability is of little consequence, but, in fact, in everyday life, bubbles play an important role (e.g., from the functioning of the lungs to the enjoyment of beer and champagne!). It is a common observation that, ordinary water, when shaken, does not form any bubbles at the surface. On the other hand, all soap and detergent solutions (shampoo, washing powder, beer, champagne, and seawater), on shaking, may form very extensive bubbles. In this chapter, the formation and stability of bubbles will be described. Further, even though one cannot see or observe the surface layer of a liquid directly, TLFs allow one to make some observations that can provide much useful information (Ivanov, 1988; Birdi, 2002).

8.2 BUBBLES AND FOAMS

If pure water is shaken, no bubbles are observed at the surface. All pure organic fluids exhibit no bubble formation on shaking. It means that, as an air bubble rises to the surface of the liquid, it merely exits into the air. However, if an aqueous detergent (surface-active substance) solution is shaken or an air bubble is created under the surface, a bubble is formed (Figure 8.1).

The formation of a bubble can be described as follows:

The process begins with an air bubble inside the liquid phase. At the surface, the bubble detaches and moves up under gravity. The detergent molecule forms a bilayer in the bubble film. The water in between is the same as the bulk solution. This may be depicted as follows: a surface layer of detergent is applied, a bubble forms with air and a layer of detergent, and the bubble at the surface forms a double layer of detergent with some water in between TLF; varying from 10 μm to 100 μm).

The bubble test of shaking a water solution, although simple, is very sensitive and may be used to determine the presence of very minute (around parts per million) contents of surface-active substances.

A bubble is composed of a TLF with two surfaces, each with a polar end pointed inward and the hydrocarbon chains pointing outward (Figure 8.2). The water inside the films will move away (due to gravity), effecting the thinning of the film. Since the

161

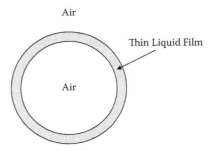

FIGURE 8.1 Formation of a bubble.

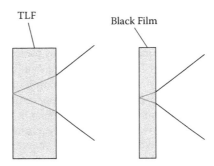

FIGURE 8.2 TLF and reflection of light.

thickness of the film approaches the dimensions of the wavelength of light, varying interference colors are observed. The reflected ray will interfere with the incident wavelength. The consequence of this will be that, depending on the thickness of the film, one will observe colors, especially when the thickness of the film is approximately the same as the wavelength of the light (i.e., 400–1000 Å). The black film is observed when the thickness is the same as the wavelength of the light (approximately 500–700 Å). Thus, this provides the closest visual observation of two-molecule thick film by eyesight.

8.2.1 APPLICATION OF BUBBLES IN TECHNOLOGY

Bubbles are critical in manufacturing in the food industry. The stability and size of the bubbles determines the taste and the looks of the product. In industry, much research has been done on the factors that control bubble formation and stability. This is of special interest in the production of ice cream, where air bubbles are trapped in frozen material.

8.3 FOAMS (THIN LIQUID FILMS)

Ordinary foams from detergent solutions are initially thick (measured in micrometers), and, as the fluid flows away, due to gravity or capillary forces or surface evaporation, the film becomes thinner (by a few hundred angstroms).

The foam consists of

Air on one side
Outer monolayer of detergent molecule
Some amount of water
Inner monolayer of detergent molecule
Air on outer layer

This can be depicted as follows (schematic):

DETERGENT_WATER_**DETERGENT**
DETERGENT_WATER_**DETERGENT** **DETERGENT**_WATER_**DETERGENT**
DETERGENTWATER**DETERGENT**

The orientation of the detergent molecule in TLF is such as that the polar group (OO) is pointing toward the water phase, and the apolar alkyl part (CCCCCCCCC) is pointing toward the air, as

    AirCCCCCCCCCCOOWATEROOCCCCCCCCCCAir
    AirCCCCCCCCCCOOWATEROOCCCCCCCCCCAir
    AirCCCCCCCCCCOOWATEROOCCCCCCCCCCAir
    AirCCCCCCCCCCOOWATEROOCCCCCCCCCCAir
    AirCCCCCCCCCCOOWATEROOCCCCCCCCCCAir
    AirCCCCCCCCCCOOWATEROOCCCCCCCCCCAir

The thickness of the water phase can vary from over 100 μm to less than 100 nm. Foams are thermodynamically unstable since there is a decrease in total free energy when they collapse. As the thickness increases around the wavelength of light (nm), one starts to observe rainbow colors (arising from the interference). The TLF forms at an even smaller thickness (50 Å or 5 nm).

However, certain kinds of foams are known to persist for very long periods of time, and many attempts have been made to explain their metastability.

The TLF may be regarded as a kind of condenser. The repulsion between the two surfactant layers (Figure 8.3) will be determined by the EDL. The effect of added ions to the solution is to make the EDL contract, and this leads to thin films.

The film looks black-gray, and its thickness is around 50 Å (5 nm), which is almost the size of the bilayer structure of the detergent (i.e., twice the length [ca. 25 Å] of a typical detergent molecule). Actually, it is a remarkable fact that one can _see_ two molecules' thin structure. The rainbow colors are observed since the light is reflected by the varying thickness of the TLF of the bubble.

In the beer industry, foaming behavior is vital to the product. The beer bottle is produced under CO_2 gas at high pressure. As soon as a beer bottle is opened, the pressure drops and the gas (CO_2) is released, which gives rise to foaming. Commonly, the foam stays inside the bottle. Foaming is caused by the presence of different amphiphilic molecules (fatty acids, lipids, and proteins). The foam is very rich as the liquid film is very thick and contains a substantial aqueous phase (such foams are

Without Salt

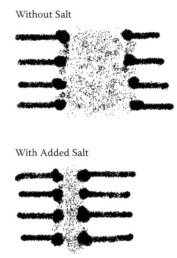

With Added Salt

FIGURE 8.3   Thickness of foam films and added ions.

called *kugelschaum* seeross). The foam fills the empty space in the bottle, and under normal conditions, it barely spills out. However, under some abnormal conditions, the foam is highly stable and starts to pour out of the bottle, which is considered undesirable. In fact, if a batch of beer shows this property, it is mixed with other batches. There are some reports that addition of heavy metal ions could change the foaming characteristics (Birdi, 1989).

Regarding foam stability, it has been recognized that the surface tension under film deformation must always change in such a way as to resist the deforming forces. Thus, tension in the film where expansion takes place will increase, while it will decrease in the part where contraction takes place. A force exists that tends to restore the original condition, which is film elasticity, defined as

$$E_{film} = 2\ \mathbf{A}\ (d\gamma/d\mathbf{A}) \tag{8.1}$$

where E_{film} and $\mathbf{A}$ are the elasticity and area of the film, and γ is the surface tension of the surface deformed.

Bubble formation in champagne is another important area of application. The size of the bubbles is found to be important for taste to make an impact. Stability has also much impact on the looks and taste. The taste of impact is related to the size and number of bubbles, and also how long the bubbles are stable.

The stability of any foam film is related to the kinetics of thinning of the TLF. As the thickness reaches a critical value, stability becomes critical. It was recognized at a very early stage by Gibbs that an unstable state of thickness will conform when the film diverges from bulk system properties. The measure was mentioned to be in the range of 50–150 Å. This state is called the *black film,* and the random motion of the molecules may easily cause a rupture of TLF. The factors determining drainage of the fluid in these films are the flow of liquid under gravity and the mean velocity not exceeding $1000 \ \mathbf{D}_{film}{}^2$, the distance between the lamella (Gibbs, assuming the fluid in film has the same viscosity and density).

D (mm)	Flow (mm/s)
0.01	0.1
0.001	0.001

Also to be considered is the suction flow due to curvature (can be much greater than the gravity effect) and the evaporation loss.

The latter will depend on the surroundings, and will be almost negligible in a closed container.

8.3.1 FOAMS STABILITY

As is known, if one blows air bubbles in pure water, no foam is formed. On the other hand, if a detergent or protein (amphiphile) is present in the system, adsorbed surfactant molecules at the interface produce foam or soap bubble. Foam can be characterized as a coarse dispersion of a gas in a liquid, where the gas is the major phase volume. The foam, or the lamina of liquid, will tend to contract due to its surface tension, and a low surface tension would thus be expected to be a necessary requirement for good foam-forming property. Furthermore, in order to be able to stabilize the lamina, it should be able to maintain slight differences of tension in its different regions. Therefore, it is also clear that a pure liquid, which has constant surface tension, cannot meet this requirement. The stability of such foams or bubbles has been related to monomolecular film structures and stability. For instance, foam stability has been shown to be related to surface elasticity or surface viscosity, η_s, besides other interfacial forces.

Foam destabilization is also a factor in the packing and orientation of mixed films, which can be determined from monolayer studies. It is worth mentioning that foam formation from monolayers of amphiphiles constitutes the most fundamental process in everyday life. The other assemblies, such as vesicles and BLM, are somewhat more complicated systems, which are also found to be in equilibrium with monolayers.

Although the surface potential, ψ, the electrical potential due to the charge on the monolayers, will clearly affect the actual pressure required to thin the lamella to any given dimension, we shall assume, for the purpose of a simple illustration, that $1/k$, the mean Debye–Huckel thickness of the ionic double layer, will influence the ultimate thickness when the liquid film is under relatively low pressure. Let us also assume that each ionic atmosphere extends only to a distance of $3/k$ into the liquid when the film is under a relatively low excess pressure from the gas in the bubbles; this value corresponds to a repulsion potential of only a few millivolts. Thus, at about 1 atm pressure:

$$\mathbf{h}_{film} = 6/k + 2 \text{ (monolayer thickness)} \tag{8.2}$$

For charged monolayers adsorbed from 10^{-3} n-sodium oleate, the final total thickness, $\mathbf{h}_{film}$, of the aqueous layer should be of the order 600 Å (i.e., $6/k$ or $18/oc$ Å).

To this value one needs to add 60 Å (=10^{-10} m) for the two films of oriented soap molecules, giving a total of 660 Å. The experimental value is 700 Å. The thickness decreases on the addition of electrolytes, as is also suggested by Equation 8.2. For instance, the value of h_{film} is 120 Å in the case of 0.1 M-NaCl. The addition of a small amount of certain nonionic surface-active agents (e.g., n-lauryl alcohol, n-decyl glycerol ether, laurylethanolamide, laurylsufanoylamide) to anionic detergent solutions has been found to stabilize the foam. It was suggested that the mode of packing is analogous to the palisade layers of the micelles and the surface layers of the foam lamellae.

Measurements have been carried out on the excess tensions, equilibrium thicknesses, and compositions of aqueous foam films stabilized by either n-decyl methyl sulfoxide or n-decyl trimethyl ammonium-decyl sulfate, and containing inorganic electrolytes.

It was recognized at a very early stage (Marangoni, 1871; Gibbs, 1878; Birdi, 2002, 2007) that the stability of a liquid film must be greatest if the surface pressure strongly resists deforming forces. In case the area of the film is extended by a shock (or vibration), then the change in surface pressure, Π, is given as

$$\Pi = -[d \ \Pi/dA] \ [A2 - A1] \qquad (8.3)$$

where A1 and A2 are, respectively, the available areas per molecule of the foam-stabilizing agent in the original and in the extended parts of the surface. This can be written as

$$\Pi = - A1 \ [d \ \Pi/dA] \ [A2/A1 - 1] \qquad (8.4)$$

$$= - A1 \ [d \ \Pi/dA] \ [j - 1] \qquad (8.5)$$

$$= Cs^{-1} \ [j - 1] \qquad (8.6)$$

where j is the area extension factor. The term ($-A[d \ \Pi/dA]$) is the surface compressibility modulus of the monolayer. For a large restoring pressure Π, this modulus should be large. In the extended region, the local reduction of the surface pressure to ($\Pi-\Pi$) results in a spreading of molecules from the adjacent parts of the monolayer to the extended region. The tension of TLFs can be measured by applying pressure and measuring the radius of curvature. Then, using the Laplace equation, the tension can be estimated.

It has been shown (Friberg, 2003; Birdi, 2002, 2008) that there exists a correlation between foam stability and the elasticity [**E**] of the film (i.e., the monolayer). In order for **E** to be large, surface excess must be large. Maximum foam stability has been reported in systems with fatty acid and alcohol concentrations well below the minimum in γ. Similar conclusions have been observed with n-$C_{12}H_{25}SO_4Na$ [SDS] + n-$C_{12}H_{25}OH$ systems that give minimum in γ versus concentration with maximum foam at the minimum point (Chattoraj and Birdi, 1984). Because of mixed monolayer formation it has been found that SDS + $C_{12}H_{25}OH$ (and some other additives) make *liquid-crystalline* structures at the surface. This leads to a stable foam (and liquid crystalline structures). In fact, in technical formulations, SDS can be deliberately used with some (less than 1%) $C_{12}H_{25}OH$ to enhance the foaming properties.

The foam drainage, surface viscosity, and bubble size distributions have been reported for different systems consisting of detergents and proteins. Foam drainage was investigated by using an incident light interference microscope technique.

The foaming of a protein solution is of theoretical interest and also has wide application in the food industry (Friberg, 2003) and to firefighting practices. Further, in the fermentation industry where foaming is undesirable, the foam is generally caused by proteins. Since mechanical defoaming is expensive due to the high power required, antifoam agents are generally used. On the other hand, antifoam agents are not desirable in some of these systems, for instance, in food products. Further, antifoam agents deteriorate gas dispersion due to increased coalescence of the bubbles. It has been known for a long time that foams are stabilized by proteins, and that these are dependent on pH and electrolytes.

High foaming capacity is explained by the stability of the gas–liquid interface due to the denaturation of proteins, especially due to their strong adsorption at the interface, which gives rise to stable monomolecular films at the interface. The foam stability is caused by film cohesion and elasticity. Further, studies have shown that the degree of foaming of bovine serum albumin (BSA) aqueous solution was investigated. The effect of electrolytes and alcohol was investigated. A good correlation was found between the adsorption kinetics and foaming properties. The effect of partial denaturation on the surface properties of ovalbumin and lysozyme has been reported. Most protein molecules exhibit increased hydrophobicity at the interface as denaturation proceeds, due to the exposure to the outer surface of the buried hydrophobic residues (in the native state). The hydrophobicity of proteins (as described in Chapter 5) has been found to give fairly good correlation to emulsion stability in food proteins. The surface tension of these proteins decreased greatly as denaturation proceeded. The emulsifying and foaming properties of proteins were remarkably improved by heat denaturation without coagulation. The emulsifying properties increased and were found to exhibit correlation with surface hydrophobicity. The protein-foaming properties increased with denaturation. The foaming power and foam stability of $C_{12}H_{25}SO_4Na$ (SDS)-ovalbumin complexes did not improve as much as with heat-denatured protein. The surface hydrophobicity showed an increase. It is thus safe to conclude that the heat- and detergent-denatured proteins are unfolded by different mechanisms. These studies are in accord with the unfolding studies carried out comparing urea or SDS unfolding by fluorescence.

8.3.2 Foam Structure

The foam as TLF has a very intriguing structure. If (1) two bubbles of the same radius come into contact with each other, this leads to (2) the formation of contact area and subsequently to (3) formation of one large bubble.

In stage 2, the energy of the system is higher than that in stage 1, since the system has formed a contact area (dAc). The energy difference between (2) and (1) is γ dAc. When the final stage is reached (3), there will be a decrease in the total area by 41%

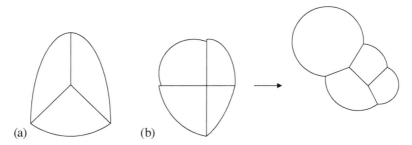

FIGURE 8.4   Foam structure of three-bubble (a) and four-bubble (b) aggregates.

(i.e., the sum of the area of two bubbles is larger than that of one bubble). This means that system 3 is at a lower energy state than the initial stage 1 (γ d AII–III).

When three bubbles come into contact, the equilibrium angle will be 120°. The angle of contact relates to system's equilibrium state. If four bubbles are attached to each other, then the angle will at equilibrium be 109° 28° (Figure 8.4).

8.3.2.1   Foam Formation of Beer and Surface Viscosity

The surface and bulk viscosities not only reduce the draining rate of the lamella but also help in restoration against mechanical, thermal, or chemical shocks. The highest foam stability is associated with appreciable surface viscosity (η_s) and yield value.

The overfoaming characteristics of beer (gushing) has been the subject of many investigations. The extreme case of gushing is when a beer, on opening, starts to foam out of the bottle and, in some cases, empties the whole bottle. The relationship between surface viscosity, η_s, and gushing was reported by various investigators. Various factors were described for the gushing process: pH, temperature, and metal ions, which could lead to protein denaturation.

The stability of a gas (i.e., N_2, CO_2, air) bubble in a solution depends on its dimensions. A bubble with a radius greater than a *critical magnitude* will continue to expand indefinitely and degassing of the solution would take place. Bubbles with a radius equal to the critical value would be in equilibrium, while bubbles with a radius less than the critical value would be able to redissolve in the bulk liquid. The magnitude of the *critical radius*, R_{cr}, varies with the degree of saturation of the liquid (i.e., the higher the level of supersaturation, the smaller the R_{cr}). The work, W, required for the formation of the bubble of radius R_{cr} is given by La Mer:

$$W = [16\,\pi\,\gamma^3]/[3\{p' - p''\}] \tag{8.7}$$

where γ is the surface tension of beer, p' is the pressure inside the bubble, p'' is the pressure in the bulk liquid. It has been suggested that there is nothing unusual in the stability of beer, and, although carbon dioxide is far from an ideal gas, empirical work supports this conclusion. A possible connection between a high value of η_s and gushing has been reported. Nickel ion, a potent inducer of gushing, has been reported to give rise to a large increase in the η_s of beer. Other additives besides Ni, such as Fe or humulinic acid, which cause gushing, have also been reported to give a large increase in η_s. On the other hand, additives that are reported to inhibit gushing, such as EDTA (ethylenediamine acetic acid—a chelating agent), have been reported to decrease the η of beer. This relation between η_s and gushing suggests that an efficient gushing inhibitor should be very surface active (in order to be able to compete with gushing promoters), but incapable of forming rigid surface layers (i.e., high η_s). Unsaturated fatty acids, such as linoleic acid, are potent gushing inhibitors since they destabilize the surface films.

The surface viscosity, η_s gram/second, was investigated by the oscillating-disc method. It was found that low η_s (0.03–0.08 g/s) beer surfaces presented a nongushing behavior. Beers with high η_s (2.3–9.0 g/s) presented a gushing behavior.

8.3.3 ANTIFOAMING AGENTS

In many cases, foaming is found to be undesirable (such as in dishwashing and wastewater treatment), and in many industries, such as in wastewater treatment, all foaming needs to be eliminated. The main criteria for antifoaming molecules have the following characteristics:

They do not form mixed monolayers.
Surface viscosity is reduced (thus destabilizing the foam films).
The boiling point is low (such as with ethanol).

8.4   APPLICATIONS OF FOAMS

8.4.1   WATER PURIFICATION

The biggest challenge mankind is facing is the need for supplying pure drinking water worldwide. The world population increase (from 1900 to 2000 by a factor 4) is much faster than the availability of clean drinking water supplies. Further, the increased need for water in industrial production also adds a further burden on water supply. The purification of water for households has been developed during the past decades, but the pollutants found in wastewater are of a different origin and concentration. Solid particles are mostly removed by filtration, but colloidal particles are not easily removed by this method. Solute compounds are rather difficult to remove, especially toxic substances with very low concentration. Flotation has been used with great advantage in some cases where sedimentation cannot remove all the suspended particles. The following are examples where flotation is being used with much success:

Paper fiber removal in the pulp and paper industries
Oils, greases, and other fats in food, oil refinery, and laundry wastes
Clarification of chemically treated waters in potable water production
Sewage sludge treatment

Many of the industrial wastewaters amenable to clarification by flotation are colloidal in nature (e.g., oil emulsions, pulp and paper wastes, and food processing). For

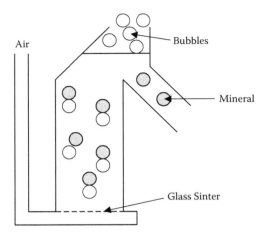

FIGURE 8.5 Apparatus used for froth flotation experiments.

the rest, wastes must be coagulated prior to flotation. In fact, flotation is always the last step in treatment. In order to aid flotation effectivity, surfactants are used. This leads to lower surface tension and foaming. The latter helps in retaining particles in the foam under flotation.

8.4.2 FROTH FLOTATION AND BUBBLE FOAM PURIFICATION

It was found many decades ago that foam or bubbles could be used to purify wastewater. A simple device used for laboratory froth flotation studies is shown in Figure 8.5.

Bubbles are formed in the sintered glass as air or other suitable gases (N_2, CO_2, etc.) is bubbled through the solution containing the solid suspension. A suitable flotation agent (a suitable-surface active agent) is added, and the air is bubbled.

Surface-active pollutants in wastewater have been removed by bubble film separation methods. Very minute concentrations are easily removed by this method, which is more economical than more complicated methods (such as active charcoal and filtration). This method is now commercially available for such small systems as fish tanks, etc. The principle in this procedure is to create bubbles in the wastewater tank and to collect the bubble foam at the top (Figure 8.6).

Bubbles are blown into the inverted funnel. Inside the funnel, the bubble film is transported away and collected. Since the bubble film consists of a surface-active substance and water, it is seen that even very minute amounts (less than milligram per liter) of surface-active substances will accumulate at the bubble surface. As shown previously, it would require a large number of bubbles to remove a gram of substance. However, since one can blow thousands of bubbles in a very short time, the method is found to be very feasible.

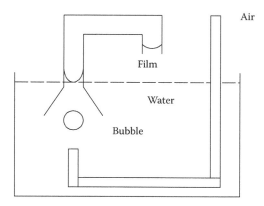

FIGURE 8.6 Bubble foam separation method for wastewater purification.

9 Emulsions, Microemulsions, and Lyotropic Liquid Crystals

9.1 INTRODUCTION

In this chapter, special application of surface chemistry principles pertaining to the oil and water phases will be considered. Oil and water *do not mix* if shaken. As is well known, if one shakes oil and water, oil breaks up into small drops (of about a few millimeters in diameter), but these drops join together rather quickly to return to their original state (as shown here).

Step I: Oil phase and water phase
Step II: Mixing
Step III: Oil drops in water phase
Step IV: After a short interval
Step V: Oil phase and water phase

However, oil and water can be dispersed with the help of suitable *emulsifiers* (surfactants) to give *emulsions* (Sjoblom et al., 2008; Birdi, 2008). This is a well-known fact with emulsions found in the home, such as mayonnaise, the basic reason being that the interfacial tension (IFT) between oil and water is around 50 mN/m, which is high, and which leads to the formation of large oil drops. On the other hand, the addition of suitable emulsifiers reduces IFT to very low values (even much less than 1 mN/m). Emulsion formation means that oil drops remain dispersed for a given length of time (even up to many years). The stability and the characteristics of these emulsions are related to the areas of their applications.

Emulsions are mixture of two (or more) immiscible substances. Everyday common examples are milk, butter (fats, water, salts), margarine, mayonnaise , skin creams, and others. In butter and margarine, the continuous phase consists of lipids. These lipids surround the water droplets (water-in-oil emulsion). All technical emulsions are prepared by some kind of mechanical agitation or mixing. Remarkably, the natural product, milk, is made by organisms without any agitation inside the mammary glands.

Emulsions are one of the most important application areas of surface-active compounds, and are generally described as belonging to three different kinds:

Emulsions
Microemulsions
Liquid crystals (LC; and lyotropic LC)

They are systems in which both water and oil need to be used in an application; it may be skin treatment, or shining of shoes or a similar activity. In other words, both these two components (water and oil), which do not mix, should be applied simultaneously. This also allows the performance of functions that are dependent on water or oil.

In most emulsion systems, there are two liquid phases that are involved, such as the water and the oil phase. There is thus a need for information regarding IFT between oil and water, as well as the solubility characteristics of surface-active substances (SAS) needed to stabilize emulsions.

Microemulsions are microstructured mixtures of oil, water, emulsifiers, and other substances. Since their structures differ in many ways from that of ordinary emulsions, it will be described separately. *Liquid crystals* (LC) are substances that exhibit special melting characteristics. Further, some surfactant–water–cosurfactant mixtures may also exhibit LC (lyotropic crystal) properties.

Thus emulsion technology is basically concerned with preparing mixtures of two immiscible substances, oil and water by adding suitable surface-active agents (such as emulgators, cosurfactants, and polymers).

9.2   EMULSIONS (OIL AND WATER)

When a surface-active substance is added to an oil–water system, the magnitude of IFT decreases from 50 mN/m to 30 (or lower [less than 1] mN/m. This leads to the observation that, on shaking, the decreased IFT of the oil–water system leads to smaller drops of the dispersed phase (oil or water). The smaller drops also lead to a more stable emulsion. Depending on the surfactant used, either an oil in water (O/W) or a water in oil (W/O) emulsion will be obtained. These experiments where oil and water, or oil and water + surfactant are shaken together, are shown in Figure 9.1.

These emulsions are all opaque since they reflect light. Some typical oil–water IFT values are given in Table 9.1.

One of the trends revealed by these data is that, in the case of alkanes, the decrease in IFT is much smaller with decrease in the alkyl chain than in the case of alcohols.

9.2.1   OIL–WATER EMULSIONS

Emulsions are one of the most important structures prepared specifically for a given application. For example, a day cream (skin cream) has characteristics and ingredients different from that of a night cream. One of the main differences between emulsions is whether oil droplets are dispersed in the water phase, or water drops are dispersed in the oil phase. This can be determined by measuring their conductivity because conductivity is higher for the O/W than for the W/O emulsion. Another useful property is that O/W will dissolve water while W/O will not, thus showing that W/O or O/W may be chosen depending on the application area. Especially in the

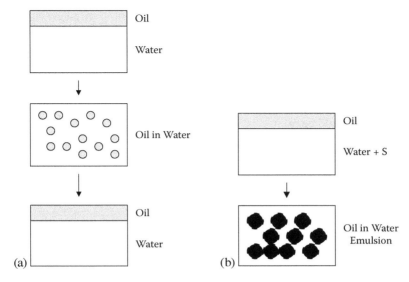

FIGURE 9.1   Mixing of oil and water (a) or oil and water + surfactant (b) by shaking.

TABLE 9.1
Magnitudes of Interfacial Tensions of Different Organic Liquids against Water (20°C)

Oil Phase	IFT (mN/m)
Hexadecane	52
Tetradecane	52
Dodecane	51
Decane	51
Octane	51
Hexane	51
Benzene	35
Toluene	36
CCl_4	45
CCl_3	32
Oleic acid	16
Octanol	9
Hexanol	7
Butanol	2

case of skin emulsions, their type and characteristics are of much importance, as in the case of day creams and night creams.

9.2.1.1 Oil-in-Water Emulsions

The main criteria for O/W emulsions is that, if water is added, it is miscible with the emulsion. Further, after water evaporates, the oil phase will be left behind. Thus, if the oil phase is needed on the substrate (such as skin, metal wood, etc.), then an O/W type emulsion should be used.

9.2.1.2 Water-in-Oil Emulsions

The criteria for W/O emulsions are that it is miscible with oil; that is, if the emulsion is added to an oil, a new but diluted W/O emulsion is obtained. In some skin creams, the W/O type emulsion is preferred, especially to obtain an oil-like feeling after application.

9.2.1.3 Hydrophilic–Lipophilic (HLB) Values of Emulsifiers

The emulsifiers commonly used exhibit varying solubility in water (or oil). This property will have consequences on the emulsion. Let us consider a system where we have oil and water. If we add an emulsifier to this system, then the latter will be distributed both in the oil and water phases. The degree of solubility in each phase will of course depend on its structure and HLB character. The emulsifiers used in making emulsions are characterized with regard to their molecular structure. For example, amphiphile molecules have HLB characteristics. Thus, each emulsifier needed for a given system (such as an O/W or a W/O emulsion) will need a specific HLB value. The data in Table 9.2 give a rough estimation of the HLB needed for a given system of emulsion. In general, it may be expected that, if the emulsifier dissolves in water, then, on adding oil, an O/W emulsion will be obtained. Conversely, if the emulsifier is soluble in oil, then, on adding water, a W/O emulsion will result.

W/O emulsions are formed by using HLB values between 3.6 and 6, thus suggesting that emulgators that are soluble in the oil phase are generally used. O/W emulsions need HLB values around 8 to 18. This is only a very general observation; it must be noted that HLB values alone do not determine emulsion type. Other parameters, such as temperature, properties of the oil phase, and electrolytes in the aqueous phase also have their effect. However, HLB values have no relation to the degree

TABLE 9.2

HLB Values of Different Emulsifiers Commonly Used in Emulsions

Emulsifier Solubility	HLB	Application in Water
Very low solubility	0–2	W/O
Low solubility	4–8	W/O
Soluble	10–12	Wetting agent
High solubility	14–18	O/W

TABLE 9.3
HLB Values of Some Typical
Surface-Active Agents

Surface-Active Agents	HLB
Na-lauryl sulfate	40
Na-oleate	18
Tween80 (sorbitan monooleate EO20)	15
Tween81 (sorbitan monooleate EO6)	10
Ca-dodecylbenzene sulfonate	9
Sorbitan monolaurate	9
Soya lecithin	0.8
Sorbitan monopalmiate	0.7
Glycerol monolaurate	5
Sorbitan monostearate	0.5
Span80 (sorbitan monooleate)	4
Glycerol monostearate	4
Glycerol monooleate	3
Sucrose distearate	3
Cetyl alcohol	1
Oleic acid	1

of emulsion stability. The HLB values of some surface-active agents are given in Table 9.3.

HLB values decrease as the solubility of the surface-active agent *decreases in water.* Solubility of cetyl alcohol in water (at 25°C) is less than a milligram per liter. It is thus obvious that, in any emulsion, cetyl alcohol will be present mainly in the oil phase, while SDS will be mainly found in the water phase. Empirical HLB values are found to have significant use in emulsion technology applications. It was shown that HLB is related, in general, to the distribution coefficient, K_D, of the emulsifier in the oil and water phases:

$$K_D = C(water)/C(oil) \qquad (9.1)$$

where C(water) and C(oil) are the equilibrium molar concentrations of the emulsifier in the water and oil phases, respectively. Based on this definition of K_D, it is found that

$$(HLB - 7) = 0.36 \ln (K_D) \qquad (9.2)$$

The relation between HLB and emulsion stability and structure could be suggested based on this thermodynamic relation. HLB values can also be estimated from the structural groups of the emulsifier (Table 9.4).

Food emulsifiers have found extensive application in industry. Therefore, they must satisfy special requirements of the food industry. For example, their toxicity

TABLE 9.4
HLB Group Numbers

Group	Group Number
Hydrophilic	
-SO$_4$Na	39
-COOH	21
-COONa	19
Sulfonate	11
Ester	7
-OH	2
Lipophilic	
-CH	0.5
-CH$_2$	0.5
-CH$_3$	0.5
-CH$_2$CH$_2$O	0.33

could be found from animal testing, which determines the amount of a substance that causes 50% (or more) of the test animals to die (lethal dosage; LD$_{50}$). For this reason, food emulsions are subject to very strict controls (Friberg, 1976).

9.2.2 Methods of Emulsion Formation

If one shakes oil and water, the oil breaks up into drops. However, they will quickly coalesce and return to the original state of two different phases. The longer one shakes, more drops reduce in size. In other words, the energy put into the system makes the drops smaller in size.

Emulsions are made based on different procedures. One is the use of mechanical agitation. Other methods are also used. The emulsion technology is very much a state-of-the-art type of industry (Sjoblom et al., 2008; Friberg, 1976; Holmberg, 2002; Birdi, 2002). Therefore, there exists a vast literature about methods used for any specific emulsion. In a simple case, an emulsion may be based on three necessary ingredients: water, oil, and emulsifier. In other words, one needs to determine in which weight proportions one need to mix these substances in order to obtain an emulsion (at a given temperature) to be stable (or maximum stability). This may be more conveniently carried out in a phase study in the triangle. The micellar region exists on the water–surfactant line (Figure 9.2).

Near the surfactant region the *crystalline* or *lamellar* phase is found. This is the region one finds in hand soaps. The ordinary hand soap is mainly the salt of fatty acid (coconut oil fatty acids or mixtures [85%] plus water [15%] and some salts. X-ray analyses have shown that the crystalline structure consists of a layer of soap separated by a water layer (with salts). The hand soap is produced by extruding under high pressure. This process aligns the lamellar crystalline structure lengthwise. If the degree of expansion versus temperature is measured, the expansion will be found

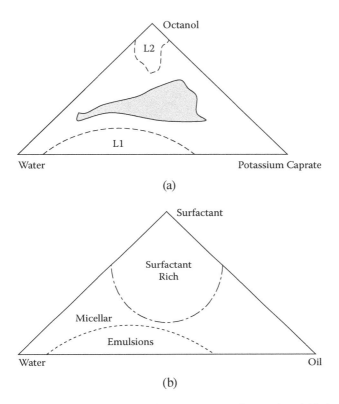

FIGURE 9.2 (a) Different phase equilibria in a water–surfactant (emulsifier)–oil mixture system (schematic). (b) Phase diagram for the system: K-caprate (PK) + water + *n*-octanol (22°C). All compositions are given in weight %. (L1 = micellar phase; L2 = reverse micelle; H1 = hexagonal LC phase [shaded]).

to be twice as much along the length than along the width. It is further seen that complex structures are present in the other regions in the phases (Figure 9.2). It is seen from the diagram that the process is strongly dependent on temperature; therefore, such studies are carried out at different temperatures.

The actual procedure is as follows. A suitable number of test samples (more than 50) are prepared by mixing each component in varying weights to represent a comparable number of regions (around 50 samples). The test samples are mixed under rotation in a thermostat over a few days to reach equilibrium. The samples are then centrifuged and the phases are analyzed. From these analyses, the phase structures are determined. They are then investigated using a suitable analytical method.

It is obvious that studies of multicomponent systems will lead to a very large numbers of phases. However, by analyzing a typical system, some trends can be found that can be used as guidelines.

A historically well-investigated system consists of water, potassium caprate (K-caprate), and *n*-octanol. The phases were determined as indicated in Figure 9.2a.

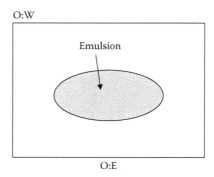

FIGURE 9.3   Emulsion region based on ratio of oil (O):water (W) versus oil (O):emulsifier (E).

This system is a very useful example for understanding what phase equilibria are involved when three components are mixed. Some characteristics are noticeable in this system that point out the significance of the ratios between KC and O. For example, the aqueous phase region is extrapolated to 1 mol octanol:2 mol K-caprate. This shows that the 1:2 ratio dominates the phase region. It has been found from other studies (such as monolayers on water films of lipids) that such mixtures are indeed found. The three-phase region is extrapolated to show that the ratio is 1 mol octanol:1 mole KC. It is found that, in such complicated phase equilibria, some simple molecular ratios indicate the phase boundaries. Thus, in general, it may safely be concluded that these *molecular ratios* will be useful when working with emulsions. The observation that exact ratios exist between different components at the phase lines suggests that some kind of molecular aggregates are formed. These correspond to the formation liquid-crystalline structures. Confirmation of these molecular aggregates has been found from monolayer studies of mixed films spread on water (Birdi, 1984, 1989). A similar conclusion was reached when investigating microemulsions (as described in the following text).

Further, in practice, a given emulsion needs to be prepared with some specified range of ratio between oil and water. In such cases, it may be more useful to study mixtures of oil (O), water (W), emulsifier (E), as plots of ratios (Figure 9.3). The region of the most suitable emulsion can be determined by studying varying mixtures.

9.2.3   EMULSION STABILITY AND ANALYSES

The stability of emulsions is dependent on various parameters such as size of drops and interactions between drops. These are described in the following text.

Emulsion drop size analyses: Since stability and other characteristics (such as viscosity and appearance) are known to be related to the drop size, they need to be measured. There are commercial instruments that can be used for such analyses. These are

Coulter counter: This is the most common type where one simply counts the number of particles or drops passing through a well-defined hole. A signal is produced, which corresponds to the size of the particle.

Light scattering: Modern laser light scattering instruments are very advanced devices for particle size distribution analyses. The laser light is scattered by the small dispersed particles or drops. The latter is known to be dependent on the radius of the particle.

9.2.3.1   Emulsion Stability

Emulsions are stable as long as the drops are separated from each other. Flocculation of an emulsion or dispersion takes place upon collision of the droplets, which is related to the Brown motion, convective stirring, or gravitational forces. Actually, any emulsion can be separated into oil and water phases by suitable centrifugation treatment.

The dispersion force of attraction between two bodies i and j (molecules, particles, drops), E_{ij}, is dependent on following parameters:

$$E_{ij} = H_{ij}/(12 \, \Pi \, \mathbf{R}^2) \tag{9.3}$$

where H_{ij} is the Hamaker constant for i and j, and $\mathbf{R}$ is the distance between the particles. Since in emulsions we have oil (1) and the continuous medium water (2), then the expression for E_{121} is found to be

$$E_{121} = (\mathbf{a} \, H_{121})/(12 \, \mathbf{R}) \tag{9.4}$$

where $\mathbf{a}$ is the size of the oil drop. The Hamaker constant, H_{121}, is found to be related to the dispersion surface tension, γLD, such that, for oil/water emulsion,

$$H_{121} = 3 \; 10^{-14}/\varepsilon_2 \; (\gamma_{LD} \, 0.5 - \gamma_{2D} \, 0.5)^{11/6} \tag{9.5}$$

where γ_{LD} (30 mN/m) and γ_{2D} (22 mN/m) are the dispersion surface tensions of oil and water, respectively. From these equations, we find that, if oil/water

$$\gamma_{LD} = 30 \text{ mN/m}; \; \varepsilon_2 = 1.77 \tag{9.6}$$

then H_{121} is equal to 1.1 10^{-14} ergs. For drops of size equal to 1 µcm (a = 5 10^{-5} cm), E_{121} is equal to almost $k_B T$ (4 10^{-14} erg, at 298 K [25°C]).

H_{121} has been shown to be always positive, which suggests that, in two-phase systems (such as oil–water), the particles will always be attracted to each other. This means that even air bubbles will attract each other, as is also found from experiments. A linear relation is found between $H_{121}{}^{6/11}$ and γ_{LD}, as expected from Equation Experimental values of A_{121} as determined from flocculation kinetics showed that this agreed with the theoretical relation.

9.2.3.2   Electrical Emulsion Stability

Those systems in which the emulsifier carries a charge would impart specific characteristics to the emulsion. A double layer will exist around the oil droplets in an O/W emulsion. If the emulsifier is negatively charged, then it will attract positive counterions while repelling negative charged ions in the water phase. The change in

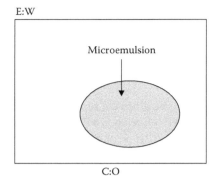

FIGURE 9.4 Four-component system: Oil (O)–Water (W)–Emulsifier (E)–Cosurfactant (S) (ratio of O:S versus S:W).

potential at the surface of oil droplets will be dependent on the concentration of ions in the surrounding water phase.

The state of stability under these conditions can be qualitatively described as follows. As two oil droplets approach each other, the negative charge gives rise to a repulsive effect (Figure 7.4). The repulsion will take place within the electrical double-layer (EDL) region. It can thus be seen that the magnitude of double-layer distance will decrease if the concentration of ions in the water phase increases. This is because the electrical double layer region decreases. However, in all such cases in which two bodies come closer, there exists two different kinds of forces that must be considered:

$$\text{Total force} = \text{repulsion forces} + \text{attraction forces}$$

The nature of the total force thus determines whether

    The two bodies will stay apart
    The two bodies will merge and form a conglomerate

This is a very simplified picture, but a more detailed analysis will be given elsewhere. The attraction force arises from van der Waals forces. The kinetic movement will finally determine whether the total force can maintain contact.

9.2.4 Orientation of Molecules at Oil–Water Interfaces

At this stage in the literature, there is no method available by which one can directly determine the orientation of molecules of liquids at interfaces. Molecules are situated at interfaces (e.g., air–liquid, liquid–liquid, and solid–liquid) under asymmetric forces. Recent studies have been carried out to obtain information about molecular orientation from surface tension studies of fluids (Birdi, 1997). It has been concluded that interfacial water molecules, in the presence of charged amphiphiles, are in a tetrahedral arrangement similar to the structure of ice. Extensive studies of alkanes near their freezing point had indicated that surface tension changes in abrupt steps. X-ray scattering of liquid surfaces indicated similar behavior (Wu et al., 1993). However, it was found that lower-chain alkanes (C_{16}) did not show this behavior. The crystallization of C_{16} at 18°C shows an abrupt change due to the contact angle change at the liquid–Pt plate interface (Birdi, 1997). It was found that, in comparison to the C_{16}–air interface, one observes supercooling (to ca. 16.4°C). Each data point corresponded to 1 s, and the data showed that crystallization is very abrupt. High-speed data (<<1 s) acquisition is needed to determine the kinetics of transition. This kinetic data would add more information about molecular dynamics at the interfaces, and the effect of additives to the aqueous phase, such as proteins. The magnitude of IFT is 12.6 and 4 mN/m for bovine serum albumin (BSA) and casein, respectively. These difference are described in detail in the literature (Birdi, 2002).

9.3 MICROEMULSIONS

As mentioned earlier, ordinary emulsions as prepared by mixing oil, water, and emulsifier are thermodynamically unstable. That is, such an emulsion may be stable over a length of time, but it will finally separate into two phases (the oil phase and the aqueous phase). They can also be separated by centrifugation. These emulsions are opaque, which means that the dispersed phase (oil or water) is present in the form of large droplets (more than a micrometer and thus visible to the naked eye).

A *microemulsion* is defined as a thermodynamically stable and clear isotropic mixture of water–oil–surfactant–cosurfactant (in most systems, it is a mixture of short-chain alcohols). The cosurfactant is the fourth component, which effects the formation of very small aggregates or drops that make the microemulsion almost clear.

Microemulsions are also characterized as microstructured, themodynamically stable mixtures of water, oil, surfactant, and additional components (such as cosurfactants). The study of microemulsions has shown that they are of the following types:

Microdroplets of oil in water or water in oil
Bicontinuous structure

Emulsifier will be found in both these phases. On the other hand, in systems with four components (Figure 9.4), consisting of oil–water–detergent–cosurfactant, there exists a region where a clear phase is found. This is the region where microemulsions are found.

Microemulsions are thermodynamically stable mixtures. The interfacial tension is almost zero. The size of drops is very small, and this makes the microemulsions look clear. It has been suggested that microemulsion may consists of bicontinuous structures, which sounds more plausible in these four-component microemulsion systems. It has also been suggested that microemulsion may be compared to swollen micelles (i.e., if one solubilizes oil in micelles). In such isotropic mixtures, short-range order exists between droplets. As found from extensive experiments, not all mixtures of water–oil–surfactant–cosurfactant produce a microemulsion. This has led to studies that have attempted to predict the molecular relationship.

Microemulsions have been formed by one of following procedures:

Oil–water mixture is added to a surfactant. To this emulsion, a short-chain alcohol (with four to six carbon atoms) is added continuously until a clear mixture (microemulsion) is obtained. Microemulsions will exhibit very special properties, quite different from those exhibited by ordinary emulsions; the microdrops may be considered as large micelles.

A very typical microemulsion, extensively investigated, consists of a mixture of

$$SDS + C_6H_6 + water + cosurfactant (C_5OH \text{ or } C_6OH)$$

The phase region is determined by adding various mixtures (approximately 20 samples) and allowing the system to reach equilibrium under controlled temperature. In the literature, the following recipe is found (Birdi, 1982):

Mix 0.0032 mol (0.92 g) SDS (mol. wt. of SDS $(C_{12}H_{25}SO_4Na) = 288$) with 0.08 mol (1.44 g) water and add 40 mL of C_6H_6. This mixture is agitated by vigorous stirring, and a creamy emulsion is obtained. To this three-component mixture, a cosurfactant ($C_5H_{11}OH$ or $C_6H_{13}OH$) is added under slow stirring until a clear system consisting of a microemulsion is obtained. The stability region is found to be a relation between surfactant–water and surfactant–alcohol. This shows that, at the molecular level, a liquid crystal structure indeed involved. The size of oil droplets is under a microgram, and therefore the mixture is clear (Birdi, 1982). These data clearly indicate that the microemulsion phase is formed at certain fixed surfactant:water and cosurfactant:oil ratios.

It is important to consider the different stages when producing microemulsions from macroemulsions. It was mentioned earlier that surfactant molecules orient with the hydrophobic group inside the oil phase, while the polar group orients toward the water phase. The orientation of surfactants at the interfaces cannot be measured by any direct method, although much useful information can be obtained from monolayer studies of the air–water or oil–water interfaces.

At present, it is generally accepted that a microemulsion recipe cannot be easily predicted. However, some suggestions have been made regarding certain criteria, which may be summarized as follows:

The HLB value of the surface needs to be determined (for deciding the O/W or W/O type).

The phase diagram of the water–oil–surfactant (and cosurfactant) needs to be determined.

The effect of temperature is found to be very crucial.

The effect of added electrolytes is of additional importance.

Recently, the phase equilibria of a microemulsion were reported. The phase behavior of a microemulsion formed with food-grade surfactant sodium bis-(2-ethylhexyl) sulfosuccinate (AOT) was studied. Critical microemulsion concentration (cµc) was deduced from the dependence of the pressure of cloud points on the concentration of

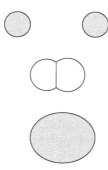

FIGURE 9.5   Two oil drops (top); flocculation step (middle); coalescence (bottom).

surfactant AOT at constant temperature and water concentration. The results show that there are transition points on the cloud point curve in a very narrow range of concentration of surfactant AOT. The transition points changed with temperature and water concentration. These phenomena show that a lower temperature is suitable for forming microemulsion droplets, and that a microemulsion with high water concentration is likely to absorb more surfactants into the structure at the interface.

9.3.1   DIFFERENT EMULSION RECIPES

9.3.1.1   Cleaning and Polishing Emulsions

In the application of emulsions for cleaning and polishing, a water phase and an oil phase (a selected halogenated solvent) are needed. The emulsions are used in cleaning and polishing the surfaces of metals and other hard surfaces (glass, etc.).

For cleaning grease from metal surfaces, a strong alkali media (soda ash, borax, or alkali phosphates or silicates) are needed. However, these alone cannot effectively remove baked-on grease from metal surfaces. However, mixtures of soda and emulsifiers are found to be effective.

9.3.1.2   Microemulsion Detergent

The following examples indicate the methodology of applying micromemulsions.

Liquid detergent formulation: A light-duty microemulsion liquid detergent composition, useful for removing greasy soils from surfaces with both concentrated and diluted forms, has been reported. It consists of the following components:

A moderately water-soluble complex of anionic and cationic surfactants (1 to 10%), in which the complex (the anionic and cationic moieties) are in essentially equivalent or equimolar proportions, an anionic detergent,

A cosurfactant (1 to 5%)
An organic solvent (1 to 5%)
Water (70%)

This recipe is based on the following considerations. It is known that if an anionic detergent (such as SDS) is mixed with a cationic detergent (such as CTAB), then a complex (1 mol:1 mol) is formed, which is sparingly soluble in water. The reason is that positively and negatively charged moieties interact and produce a neutral complex (which is insoluble in water). This complex is oil soluble.

The complex component is one in which the anionic and cationic moieties include hydrophilic portions or substituents in addition to the complex-forming portions thereof. The anionic detergent is a mixture of higher paraffin sulfonate and higher alkyl polyoxyethylene sulfate. The cosurfactant is a polypropylene glycol ether, a poly-lower-alkylene glycol lower-alkyl ether, or a poly-lower-alkylene glycol lower alkanoyl ester, and the organic solvent is a nonpolar oil, such as an isoparaffin, or an oil having polar properties, such as a lower fatty alkyl chain. Such a light-duty, clear microemulsion liquid detergent composition has been reported to be useful for the removal of greasy soils from substrates, in both concentrated form and in a diluted form with water.

9.3.2 Characteristics and Stability of Emulsions

The stability of any emulsion is dependent on needs and the application area. In some cases, the emulsion need to be *stable* for longer time than in other cases. As in the case of hair cream, the emulsion should *destabilize* as soon as it is applied to the hair, as otherwise, the hair will be white with emulsion droplets. On the other hand, any emulsion used in spraying on plants needs to be stable for longer time. Further, if one needs to clean oil spills on oceans, the emulsion formation then needs to be *destabilized*.

There are different processes that are involved in the stability and characteristics of emulsions.

They are as follows:

Creaming or flocculation of drops: This process is described in those cases where oil drops (for oil–water) cling to each other and grow in large clusters. The drops do not merge into each other. The density of most oils is lower than that of water. This leads to instability as the oil drop clusters rise to the surface (Figure 9.5).

This process can be reduced by
1. Increasing the viscosity of the water phase and thereby decreasing the rate of movement of the oil drops
2. Decreasing interfacial tension and thus the size of the oil drops

The ionized surfactants will stabilize O/W emulsions by imparting the surface electrical double layer (**EDL**).

9.4 LIQUID CRYSTALS

Liquid crystals (LC) are phase structures that are intermediate between *liquid* and *crystal* phases. They have also been mentioned as mesophases (Greek *mesos* = middle). Liquid crystals have an intermediate range of order between liquid and crystal phases (Soltis et al., 2004; Friberg, 1976). LC may be described as follows. If a pure substance, such as stearic acid, is heated, it melts at a very specific temperature. Heating a pure solid shows the following behavior:

Solid ... *Melting Point* ... Liquid

However, if an LC substance is heated, it will show more than *one* melting point. Thus, liquid crystals are substances that exhibit a phase of matter that has properties between those of a conventional liquid and a solid crystal. For instance, an LC may flow like a liquid but have the molecules in the liquid arranged and/or oriented in a crystal-like way. There are many different types of LC phases that can be distinguished based on their different optical properties (such as birefringence). When viewed under a microscope using a polarized light source, different liquid crystal phases will appear to have a distinct texture. Each "patch" in the texture corresponds to a domain where the LC molecules are oriented in a different direction. Within a domain, however, the molecules are well ordered. Liquid crystal materials may not always be in an LC phase (just as water is not always in the liquid phase; it may also be found in the solid or gas phase).

Liquid crystals can be divided into the following types:

Thermotropic LCs
Lyotropic

Thermotropic LCs exhibit the phase transition into the LC phase as temperature is changed, whereas **lyotropic LCs** exhibit it as a function of concentration of the mesogen in a solvent (typically water) as well as temperature.

9.4.1 Liquid Crystal Phases

The various LC phases (called mesophases) can be characterized by the type of molecular ordering that is present. One is positional order (whether or not molecules are arranged in any sort of ordered lattice), and another is orientational order (whether or not molecules are mostly pointing in the same direction); moreover, order can be either short-range (only between molecules close to each other) or long-range (extending to larger, sometimes macroscopic, dimensions). Most thermotropic LCs will have an isotropic phase at high temperatures. That is, heating will induce in them a conventional liquid phase characterized by random and isotropic molecular ordering (a little to no long-range order) and fluid-like flow behavior. Under other conditions (for instance, lower temperature), an LC might inhabit one or more phases with significant anisotropic orientational structure and long-range orientational order while still having the ability to flow.

The ordering of liquid crystalline phases is extensive on the molecular scale. In fact, the self-assembly characteristic as possessed by lipids (amphiphiles) is the basic building feature in LCs. This order extends up to the entire domain size, which may be on the order of micrometers (μm), but usually does not extend to the macroscopic scale as often occurs in classical crystalline solids. However, some techniques (such as the use of boundaries or an applied electric field) can be used to enforce a single ordered domain in a macroscopic LC sample. The ordering in an LC might extend along only one dimension, with the material being essentially disordered in the other two directions.

9.4.1.1 Thermotropic Liquid Crystals

Thermotropic phases are those that occur in a certain temperature range. If the temperature is raised too high, thermal motion will destroy the delicate cooperative ordering of the LC phase, pushing the material into a conventional isotropic liquid phase. At too low a temperature, most LC materials will form a conventional (though anisotropic) crystal. Many thermotropic LCs exhibit a variety of phases as the temperature is changed. For instance, a particular mesogen may exhibit various smectic and nematic (and finally isotropic) phases as temperature is increased.

9.4.1.1.1 The Nematic Phase

One of the most common LC phases is the nematic, where the molecules have no positional order but no long-range orientational order either. So, the molecules flow, and their center of mass positions are randomly distributed as in a liquid, but they all point in the same direction (within each domain). Most nematics are uniaxial: they have one axis that is longer and preferred, with the other two being equivalent (can be approximated as cylinders). Some liquid crystals are biaxial nematics, meaning that, in addition to orienting their long axis, they also orient along a secondary axis.

Liquid crystals are a phase of matter whose order is intermediate between that of a liquid and that of a crystal. The molecules are typically rod-shaped organic moieties about 25 Å (2.5 nm) in length, and their ordering is a function of temperature. The nematic phase, for example, is characterized by the orientational order of the constituent molecules. The molecular orientation (and hence the material's optical properties) can be controlled with applied electric fields. Nematics are (still) the most commonly used phase in liquid crystal displays (LCDs), with many such devices using twisted nematic geometry. The smectic phases, which are found at temperatures lower than the nematic, form well-defined layers that can slide over one another like soap. The smectics are thus positionally ordered along one direction. In the Smectic A phase, the molecules are oriented along the layer normal, while in the Smectic C phase they are tilted away from the layer normal. These phases, which are liquid-like within the layers, are illustrated in the following text. There is a very large number of different smectic phases, all characterized by different types and degrees of positional and orientational order.

9.4.1.1.1.1 Chiral Phases   The chiral nematic phase exhibits chirality (*handedness*). This phase is often called the cholesteric phase because it was first observed for cholesterol derivatives. Only chiral molecules (i.e., those that lack inversion symmetry) can give rise to such a phase. This phase exhibits a twisting of the molecules along the director, with the molecular axis perpendicular to the director. The finite twist angle between adjacent molecules is due to their asymmetric packing, which results in longer-range chiral order. In the smectic phase, the molecules orient roughly along the director, with a finite tilt angle, and a twist relative to other mesogens.

The chiral pitch refers to the distance (along the director) over which the mesogens undergo a full 360° twist (the structure repeats itself every half pitch since the positive and negative directions along the director are equivalent). The pitch may be varied by adjusting temperature or adding other molecules to the LC fluid. For many types of LC, the pitch is on the same order as the wavelength of visible light, which causes these systems to exhibit unique *optical properties*, such as selective reflection.

9.4.1.2 Lyotropic Liquid Crystals

As compared to the cholesteric LC, the lyotropic LC consists of two or more components that exhibit liquid-crystalline properties (dependent on concentration, temperature, and pressure). In the lyotropic phases, *solvent molecules* fill the space around the compounds (such as soaps) to provide fluidity to the system. In contrast to thermotropic liquid crystals, these lyotropics have another degree of freedom of concentration that enables them to induce a variety of different phases. A typical lyotropic liquid crystal is surfactant–water–long-chain alcohol.

A compound that has two immiscible hydrophilic and hydrophobic parts within the same molecule is called an *amphiphilic* molecule (as mentioned earlier). Many amphiphilic molecules show lyotropic liquid-crystalline phase sequences, depending on the volume balances between the hydrophilic part and the hydrophobic part. These structures are formed through the microphase segregation of two incompatible components on a nanometer scale. Hand soap is an everyday example of a lyotropic liquid crystal (80% soap + 20% water).

The content of water or other solvent molecules changes the self-assembled structures. At very low amphiphile concentration, the molecules will be dispersed randomly without any ordering. At slightly higher (but still low) concentration, amphiphilic molecules will spontaneously assemble into micelles or vesicles. This is done so as to "hide" the hydrophobic tail of the amphiphile inside the micelle core, exposing a hydrophilic (water-soluble) surface to aqueous solution. However, these spherical objects do not order themselves in solution. At higher concentration, the assemblies will become ordered. A typical phase is a hexagonal columnar phase, in which the amphiphiles form long cylinders (again with a hydrophilic surface) that arrange themselves into a roughly hexagonal lattice. This is called the *middle* soap phase. At still higher concentration, a *lamellar* phase (neat soap phase) may form, wherein extended sheets of amphiphiles are separated by thin layers of water. For some systems, a *cubic* (also called *viscous isotropic*) phase may exist between the hexagonal and lamellar phases, wherein spheres are formed that create a dense cubic lattice. These spheres may also be connected to one another, forming a bicontinuous cubic phase.

The aggregates created by amphiphiles are usually spherical (as in the case of micelles), but may also be disc-like (bicelles), rodlike, or biaxial (all three micelle axes are distinct) (Zana, 2008). These anisotropic self-assembled nanostructures can then order themselves in much the same way as liquid crystals do, forming large-scale versions of all the thermotropic phases (such as a nematic phase of rod-shaped micelles).

For some systems, at high concentration, inverse phases are observed. That is, one may generate an inverse hexagonal columnar phase (columns of water encapsulated by amphiphiles), or an inverse micellar phase (a bulk LC sample with spherical water cavities).

A generic progression of phases, going from low to high amphiphile concentration, is

Discontinuous cubic phase (micellar phase)
Hexagonal columnar phase (middle phase)
Bicontinuous cubic phase
Lamellar phase
Bicontinuous cubic phase
Reverse hexagonal columnar phase
Inverse cubic phase (inverse micellar phase)

These structures are extensively described in the current literature (Fanum, 2008; Friberg, 1976; Birdi, 2002; Holmberg, 2004; Somasundaran, 2006). Even within the same phases, their self-assembled structures are tunable by the concentration: for example, in lamellar phases, the layer distances increase with the solvent volume. Lamellar structures are found in systems such as the common hand soap, which consists of ca. 0% soap + 20% water. The layers of soap molecules are separated by a region of water (including, salts etc.) as a kind of sandwich. The x-ray diffraction analysis shows this structure very clearly. Since lyotropic liquid crystals rely on a subtle balance of intermolecular interactions, it is more difficult to analyze their structures and properties than those of thermotropic liquid crystals. Similar phases and characteristics can be observed in immiscible diblock copolymers.

As mentioned earlier, surfactants aggregate to form micelles, which may vary in size (i.e., number of monomers per micelle) from a few to over a thousand monomers. However, surfactants can form, besides simple micellar aggregates (i.e., spherical or ellipsoidal), many other structures also when mixed with other substances. The *curved* micelle aggregates are known to change to planar interfaces when additives, the so-called cosurfactants, are added. A reported recipe consists of

Surfactant (cetylpyridinium bromide [CPBr]) + hexanol + salt + water

The addition of salts to micelles gives large micelles that turn into cylindrical shapes. However, the addition of cosurfactant produces the *liquid crystal phase*. As a consequence, these micellar systems with added cosurfactant are found to undergo several macroscopic phase transitions in dilute solutions. These transitions are as follows:

Surfactant Micelles (Spherical ... Cylinderical)

Surfactants + Cosurfactant ... Liquid crystal

Further, the extensive change of ionic charges (if present) have been found to be very prominent. The simple picture that bilayers in aqueous medium are principally stabilized by the competition between **hydration forces**–*van der Waals*–**electrostatic interactions** seem to be the most plausible basis.

The hydration forces arise from the hydrogen bond formation between the polar groups of the surfactant and the water molecule. Van der Waals forces are attraction forces between all molecules (these are short range).

Metallotropic liquid crystals: LC phases can also be based on low-melting inorganic phases such as $ZnCl_2$ that have a structure formed of linked tetrahedra and easily form glasses. The addition of long-chain soap-like molecules leads to a series of new phases that show a variety of liquid crystalline behavior both as a function of the inorganic–organic composition ratio and temperature. This class of materials has been named metallotropic.

9.4.1.3 Biological Liquid Crystals

The effect of temperature on biological living systems needs a few remarks. The human body temperature is made up of molecules and cell structures so as to function at optimum at 37°C. If the temperature changes a few degrees (plus or minus) from 37°C, then the effect is considerable but not lethal. This is because the system shows LC behavior. In other words, biological reactions can go on functioning (though with some restrictions) even if the temperature is 36°C or 38°C.

Lyotropic liquid–crystalline nanostructures are abundant in living systems. Accordingly, lyotropic LC have been of much interest in such fields as biomimetic chemistry. In fact, biological membranes and cell membranes are a form of LC. Their constituent rod-like molecules (e.g., phospholipids) are organized perpendicularly to the membrane surface; yet, the membrane is fluid and elastic. The constituent molecules can flow in plane quite easily but tend not to leave the membrane, and can flip from one side of the membrane to the other with some difficulty. These LC membrane phases can also host important proteins such as receptors freely "floating" inside, or partly outside, the membrane.

Many other biological structures exhibit LC behavior. For instance, the concentrated protein solution that is extruded by a spider to generate silk is actually an LC phase. The precise ordering of molecules in silk is critical to its renowned strength. DNA and many polypeptides can also form LC phases. Since biological mesogens are usually chiral, chirality often plays a role in these phases.

9.4.1.3.1 Theory of Liquid Crystals Formation

Microscopic theoretical treatment of fluid phases can become quite involved owing to the high material density, which means that strong interactions, hard-core repulsions, and many-body correlations cannot be ignored. In the case of LC, anisotropy in all of these interactions further complicates analysis. There are a number of fairly simple theories, however, that can at least predict the general behavior of the phase transitions in LC systems.

Additionally, the description of LC involves an analysis of order. In particular, a sharp drop of the order parameter to B is observed when a transition takes place from the LC phase into the isotropic phase. The order parameter can be measured experimentally in a number of ways, such as diamagnetism, birefringence, Raman scattering, NMR, and EPR.

A very simple model that predicts lyotropic phase transitions is the hard-rod model proposed by Onsager (Friberg, 1976). This theory considers the volume excluded from the center-of-mass of one idealized cylinder as it approaches another. Specifically, if the cylinders are oriented parallel to one another, there is very little volume that is excluded from the center-of-mass of the approaching cylinder (it can come quite close to the other cylinder). If, however, the cylinders are at some angle to one another, then there is a large volume surrounding the cylinder where the approaching cylinder's center-of-mass cannot enter (due to the hard-rod repulsion between the two idealized objects). Thus, this angular arrangement sees a decrease in the net positional entropy of the approaching cylinder (there are fewer states available to it).

It may be asserted that the fundamental reason arises from the fact that, while parallel arrangements of anisotropic objects lead to a decrease in orientational entropy, there is an increase in positional entropy. Thus, in some cases, greater positional order will be entropically favorable. This theory therefore predicts that a solution of rod-shaped objects will undergo a phase transition at sufficient concentration into a nematic phase. Recently, this theory has been used to observe the phase transition between nematic and smectic-A at very high concentration (Hanif et al.). Although this model is conceptually helpful, its mathematical formulation makes several assumptions that limit its applicability to real systems.

This statistical theory includes contributions from an attractive intermolecular potential. The anisotropic attraction stabilizes the parallel alignment of neighboring molecules, and the theory then considers a mean-field average of the interaction. Solved self-consistently, this theory predicts thermotropic phase transitions consistent with the experiment.

According to one theoretical model, it can be assumed that the liquid crystal material is a continuum. It has been suggested that three types of distortions can take place in these structures:

1. Twists of the material, where neighboring molecules are forced to be angled with respect to one another
2. Nonlinearity (or splay) of the material, where bending occurs perpendicular to the director
3. Bend of the material, where the distortion is parallel to the director and mesogen axis

All three types of distortions incur an energy penalty, and are defects that often occur near domain walls or boundaries of the enclosing container. The response of the material can then be decomposed into terms based on the elastic constants corresponding to the three types of distortions.

It has been mentioned in the literature that chiral mesogens usually produce chiral mesophases. For molecular mesogens, this means that the molecule must possess some form of asymmetry, usually a stereogenic center. An additional requirement is that the system not be racemic; a mixture of right- and left-handed versions of the mesogen will cancel the chiral effect. Because of the cooperative nature of LC ordering, however, a small amount of chiral dopant in an otherwise achiral mesophase is often enough to select out one domain handedness, making the system overall chiral.

Chiral phases usually have a helical twisting of the mesogens. If the pitch of this twist is on the order of the wavelength of visible light, then interesting optical interference effects can be observed. The chiral twisting that occurs in chiral LC phases also makes the system respond differently to right- and left-handed circularly polarized light. These materials can therefore be used as polarization filters.

It is possible for chiral mesogens to produce essentially achiral mesophases. For instance, in certain ranges of concentration and molecular weight, DNA will form an achiral line hexatic phase. A curious recent observation is of the formation of chiral mesophases from achiral mesogens. Specifically, bent-core molecules (sometimes called *banana* LCs) have been shown to form liquid crystal phases that are chiral. In any particular sample, various domains will have opposite handedness, but within any given domain, strong chiral ordering will be present.

9.4.1.3.2 *Industrial Applications of Liquid Crystals*

LCs find wide use in displays, which rely on the optical properties of certain liquid crystalline molecules in the presence or absence of an electric field. In a typical device, an LC layer sits between two polarizers that are crossed (oriented at 90° to one another). The liquid crystal is chosen so that its relaxed phase is a twisted one. This twisted phase reorients light that has passed through the first polarizer, allowing it to be transmitted through the second polarizer and reflected back to the observer. The device thus appears clear. When an electric field is applied to the LC layer, all the mesogens align (and are no longer twisting). In this aligned state, the mesogens do not reorient light, so the light polarized at the first polarizer is absorbed at the second polarizer, and the entire device appears dark. In this way, the electric field can be used to make a pixel switch between clear or dark on command. Color LCD systems use the same technique, with color filters used to generate red, green, and blue pixels. Similar principles can be used to make other LC-based optical devices.

Thermotropic chiral LCs whose pitch vary strongly with temperature can be used as crude thermometers since the color of the material will change as the pitch is changed. LC color transitions are used on many aquarium and pool thermometers. Other LC materials change color when stretched or stressed. Thus, LC sheets are often used in industry to look for hot spots, map heat flow, measure stress distribution patterns, etc. The LC in fluid form is used to detect electrically generated hot spots for failure analysis in the semiconductor industry. LC memory units with extensive capacity were used in Space Shuttle navigation equipment.

It is also worth noting that many common solutions are in fact LCs. Soap, for instance, is an LC, and forms a variety of phases, depending on its concentration in water (Friberg, 1976; Somasundaran, 2006).

9.5 APPLICATIONS OF EMULSIONS

The area of emulsion applications is very large and cannot be described in the space available here. However, some important areas will be highlighted to indicate the basic criteria for emulsion technology (Friberg, 1976; Gitis and Sivamani, 2004).

9.5.1 PERSONAL CARE INDUSTRY

Human *skin* is the outer layer of the body, that covers and protects it from any exposure to the surroundings (wind or rain, etc.). The natural substances that compose the skin are very elaborate and complex. Further, the composition of the skin changes with age, and is different for different people. This also true for the animal world.

The skin maintains communication with the underlying part of the body through pores. A transport of various substances is continuously taking place (such as water, salts, fats, etc.) through these pores. Thus, an equilibrium exists between these outer layer and the inside body. However, in some cases, this equilibrium may not be optimum, which leads to skin irritation (red skin, or rashes, or itching). This condition requires treatment of the skin using cosmetics that are suitable for curing the skin problem. At present, the personal care industry, which produces cosmetics and their products, is one of the largest worldwide.

The skin barrier properties and effect of hand hygiene practices are known to be important in protecting the body. The average adult has a skin area of about 1.75 m^2. The superficial part of the skin, the *epidermis*, has five layers. The *stratum corneum*, the outermost layer, is composed of flattened dead cells (corneocytes or squames) attached to each other to form a tough, horny layer of keratin mixed with several lipids, which help maintain the hydration, pliability, and barrier effectiveness of the skin. This part of skin has been compared to a wall of bricks (corneocytes) and mortar (lipids) and serves as the primary protective barrier. Approximately 15 layers make up the stratum corneum, which is completely replaced every 2 weeks; a new layer is formed almost daily. From healthy skin, approximately 10^7 particles are disseminated into the air each day, and 10% of these skin squames contain viable bacteria. This is a source of major dirt inside the house and contributes to many interactions.

Besides covering and protecting the body, the other function of the skin are regulation of the body temperature (ca. 37°C) and also controlling penetration of the body by sunlight, liquids, solid materials, etc. The human skin is composed of four layers of different tissue:

The underlying layer or subcutaneous tissue
The dermis or true skin

In the latter region, latter sensory nerves, blood vessels, and sweat glands are located. Here, a fatty substance called sebum is produced, which coats, lubricates, and keeps other molecules from passing through the skin. This also has an effect on the water loss from the body. If for some reason, the outer part of skin is damaged, then special creams and ointments to repair it are available commercially. These are described in the following subsection.

9.5.1.1 Fundamentals of Skin Creams and Recipes

The human skin is a rather complex structure, which protects the body against the environment. The aim of the different cosmetics is to repair and restore the original balance of elements in skin structure. Cosmetic preparations would need to consider any reaction between them and the components of the skin. Skin creams are known to be composed of a variety of ingredients, which are based on end use (hands, feet, face, hair, etc.), and some speciality products that are applied to the skin to repair effects such as dryness. We will now consider some recipes for skin creams. Since the number of personal care emulsion creams is very large, only a few typical examples are given here. A variety of emulsion skin care products are found commercially that claim to exhibit properties for nurturing and protecting the skin.

Some cosmetics have been found to contain proteins and can have moisturizing and tightening effects. A typical recipe for such purposes is as follows:

Skin cream for face (dry skin):
    Water
    Safflower seed oil
    Coconut oil
    Glycerin polysorbate 60
    Evening primrose oil
    Sorbitan stearate
    Cetyl ester
EDTA

Many procedures have been suggested for the assessment of dry skin. Evaporation of emulsion water, electrical capacitance, and skin surface (emulsion) lipids are methods used for the assessment of the efficacy of skin care products. Hygiene of the skin is one of the most important aspects of daily life, and for humans, hygiene of the skin is related to clean state. One may ask the question *how clean is clean*. Skin hygiene, particularly of the hands, is a primary mechanism for reducing contact and fecal–oral transmission of infectious agents. For over a century, skin hygiene, particularly of the hands, has been accepted as a primary mechanism to control the spread of infectious agents. However, widespread use of antimicrobial products has prompted concern about emergence of resistance to antiseptics and damage to the skin barrier associated with frequent washing. Although the causal link between contaminated hands and infectious disease transmission is one of the best-documented phenomena in clinical science, several factors have recently prompted a reassessment of skin hygiene and its effective practice.

In industrialized countries, exposure to potential infectious risks has increased because of changing sociologic patterns (e.g., more frequent consumption of commercially prepared food and expanded childcare services). Environmental sanitation and public health services, despite room for improvement, are generally good. In addition, choices of hygienic skin care products have never been more numerous, and the public has increasing access to health- and product-related information.

It has been reported that there is evidence for the relationship between skin hygiene and infection, the effects of washing on skin integrity, and recommendations for skin care practices for the public and healthcare professionals. Skin cleansing is accepted to reduce the risk of infection, and personal bathing and washing have been known for ages to be essential to maintaining good health. Though handwashing has been reported to be important for infection control, aside from hand cleansing, specific evidence is lacking to link bathing or general skin cleansing with preventing infections.

In a recent study, it was found that the use of antimicrobial soaps was associated with substantial reductions in rates of superficial cutaneous infections. Extensive experimental studies demonstrated a reduction in bacteria on the skin with use of antimicrobial soaps, but none assessed rates of infection as an outcome.

For surgical or other high-risk patients, showering with antiseptic agents has been tested for its effect on postoperative wound infection rates. Such agents, unlike plain soaps, reduce microbial counts on the skin. Whole-body washing with chlorhexidine-containing detergent has been shown to reduce infections among neonates, but concerns about absorption and safety preclude this as a routine practice. These factors have led to suggestions that antimicrobial products should be more universally used, and a myriad of antimicrobial soaps and skin care products have become commercially available. While antimicrobial-drug-containing products are superior to plain soaps for reducing both transient pathogens and colonizing flora, widespread use of these agents has raised concerns about the emergence of bacterial strains resistant to antiseptic ingredients such as triclosan.

Some evidence indicates that long-term use of topical antimicrobial agents may alter skin flora. Water content, humidity, pH, intracellular lipids, and rates of shedding help retain the protective barrier properties of the skin. When the barrier is compromised (e.g., by hand hygiene practices such as scrubbing), skin dryness, irritation, cracking, and other problems may result. Although the palmar surface of the hand has twice as many cell layers and the cells are >30 times thicker than on the rest of the skin, palms are quite permeable to water.

Damaged skin more often harbors increased numbers of pathogens. Moreover, washing damaged skin is less effective at reducing numbers of bacteria than washing normal skin, and numbers of organisms shed from damaged skin are often higher than from healthy skin. The microbial flora on the clean hands of nurses (samples taken immediately after handwashing) have been reported in several recent studies. Methicillin resistance among coagulase-negative staphylococcal flora on hands did not seem to increase during the 1980s to the 1990s, and tetracycline resistance decreased.

9.5.1.2 Detergent-Based Antiseptics or Alcohol

Because of increasingly vulnerable patient populations, the demand for hand hygiene among healthcare professionals has never been greater. However, frequent handwashing is not only potentially damaging to skin but also time consuming and expensive. A mild emulsion cleansing rather than handwashing with liquid soap was associated with a substantial improvement in the skin of nurses' hands.

Use of lotions and moisturizers: Moisturizing is beneficial to skin health and reducing microbial dispersion from the skin, regardless of whether the product used contains an antibacterial ingredient. Because of differences in the content and formulations of lotions and creams, products vary greatly in their effectiveness. Lotions used with products containing chlorhexidine gluconate must be carefully selected to avoid neutralization by anionic surfactants. The role of emollients and moisturizers in improving skin health and reducing microbial spread is an area for additional research.

For improving the skin condition of healthcare professionals and reducing their chances of harboring and shedding microorganisms from the skin, the following measures are recommended:

1. For damaged skin, mild, nonantimicrobial skin-cleansing products may be used to remove dirt and debris. If antimicrobial action is needed (e.g.,

before invasive procedures or handling of highly susceptible patients), a waterless, alcohol-based product may be used.

2. In clinical areas such as the operating room and neonatal and transplant units, shorter, less traumatic washing regimens may be used instead of lengthy scrub protocols with brushes or other harsh mechanical action.

3. Effective skin emollients or barrier creams may be used in skin care regimens and procedures for staff (and possibly patients as well).

4. Skin moisturizing products should be carefully assessed for compatibility with any topical antimicrobial products being used and for physiological effects on the skin.

A strong influence of preoperative showers on staphylococcal skin colonization has been reported for antiseptic skin cleansers. A comparison of the effects of preoperative whole-body bathing with detergent-alone and with detergent-containing chlorhexidine gluconate on the frequency of wound infections after clean surgery, comparison of preoperative bathing with chlorhexidine detergent and nonmedicated soap in the prevention of wound infection, prevention of dryness and eczema of the hands of hospital staff by emulsion cleansing instead of washing with soap, and measurements by transmission electron microscopy of "dry" skin before and after application of a moisturizing cream have all been carried out. A correlation between pH and the irritant effects of cleansers marketed for dry skin has been investigated. A novel sunscreen system based on tocopherol acetate incorporated into solid lipid *nanoparticles* has also been reported, as well as various LC applications in skin care cosmetics.

Intercellular lipids of the stratum corneum contribute threefold to the maintenance of a healthy skin: by hydration, cell adhesion, and reduction of transepidermal water loss. All of these functions can be attributed to the self-assembly property of the amphiphilic molecules of the stratum corneum lipids. A new type of skin care product called Lamellar Gel was developed, which contains a (synthesized) pseudoceramide. Its structure is similar to that of ceramide found among the stratum corneum lipids, which allows it to be compatible with intramolecular interactions. Compared to regular emulsions, the Lamellar Gel demonstrated better skin care characteristics regarding permeability, skin hydration, and skin occlusion. This was attributed to the fact that it formed the same self-organizing structure as natural stratum corneum lipids, hence showing a high affinity to the skin. A high moisturizing effect was observed as Lamellar Gel combines the benefits of both O/W and W/O emulsions: it provides the same initial hydration as an O/W emulsion and, at the same time, the same occlusivity as an W/O emulsion. Transepidermal water loss increases under dry environmental conditions. This especially affects the skin around the eyes, where the skin is very thin, and wrinkles are very easily formed. Treatment with Lamellar Gel removed these wrinkles promptly and hydrated the stratum corneum for a long time.

A novel sunscreen system based on tocopherol acetate incorporated into solid lipid nanoparticles has been developed. In recent years, solid lipid nanoparticles (SLN) have been introduced as a novel carrier system for drugs and cosmetics. It has been found that SLN possess characteristics of physical UV-blockers on their own, thus offering the possibility of developing a more effective sunscreen system with reduced side effects. Incorporation of the chemical sunscreen tocopherol acetate into SLN prevents chemical degradation and increases the UV-blocking capacity. Aqueous SLN dispersions were produced and incorporated into gels, followed by particle-size examination, stability testing upon storage, and thermoanalytical examination. Investigation of the UV-blocking capacity using different in vitro techniques revealed that the SLN dispersions produced in this study are at least twice as effective as their reference emulsions (conventional emulsions with identical lipid content). Placebo SLN show even greater UV-blocking efficacy than emulsions containing tocopherol acetate as the molecular sunscreen. Incorporation of tocopherol acetate into SLN leads to an overadditive UV-blocking effect. Furthermore, film formation of SLN on the skin and occlusivity were examined. The obtained data show that incorporation of tocopherol acetate into SLN leads to an improved sunscreen and skin care formulation.

The influence of a cream containing 20% glycerin and its vehicle on skin barrier properties has been investigated. Recent studies have shown that polymers offer several advantages and can be used in skin care products. Phase diagrams were determined for lactic and isohexanoic hydroxy acids as well as salicylic acid with water, a nonionic surfactant and a paraffinic oil, to outline the influence of hydroxy acids on the structure in a model for a skin lotion. The results showed the influence of the acid to be similar to that of the oil but that the difference in chain length between the two alpha acids had only insignificant influence. The results are discussed from two aspects: the structures involved in the lotion as applied, and the action of the lotion residue on the skin after the evaporation of the water.

In the pharmaceutics literature, an application-triggered drug release from an O/W emulsion recipe has been reported. The application of emulsions in the pharmaceutical industry is very important, and a large number of references are found in the current literature.

9.5.2   THE PAINT INDUSTRY AND COLLOIDAL ASPECTS

Everybody appreciates the function of paint is manyfold:

  Appearance
  Color
  Protection (from corrosion iron, etc.; from deterioration of wood)

Paint is recognized as a pigmented mixture (in liquid or paste form) that protects and enhances the appearance of materials. Its role in the case of cars or houses or wood furniture is well recognized. In the early days (almost 100 years ago), cars were painted with nitocellulose lacquer. Today, especially, in some areas such as cars manufacturing, the industrial development of paint has been very extensive. There are various types of paint industries:

  Paint formulation
  Paint pigments
  Paint rheology and stability

Paint stripper
Paint pretreatment
Polyester paint
Polyurethane coating
Metallizing
Marine paint and corrosion
Paint removal

Paint consists of essentially the following components:

Pigment—metallic
Binder
Dispersants (emulsifiers, etc.)
Solvent
Additives

These mixtures are very complex systems, and much of the industrial development in these areas is safeguarded by patents.

9.5.3 FOOD EMULSIONS (MILK INDUSTRY)

Surface and colloids have been found to play a big role in the processes of natural systems such as food emulsions. One of the most important functions in the life of mammals on earth is the feeding of newborns. In this context, *milk* is one of the most amazing examples of emulsions as found in nature, especially when its role in the growth and nutrition of newborns is considered. Milk consists of fats, nonfats (such as lactose), and inorganic salts. The milk emulsion has to be stable to a certain extent, and it has to provide nutrition based on the many different molecules present in the emulsion (water, fats, proteins, inorganic salts, and organic molecules). Nature uses emulsion since it needs to provide fats and proteins as in this form newborns need for their nutrition and other needs, as well as to provide water and inorganic salts. This a very remarkable feat that nature performs, by which newborns can sustain life from birth. However, in some cases, when the newborn may need extra nutrition, it has to be met by synthetic milk, a food emulsion created by the milk industry.

Another example of a food emulsion is the ice cream, in which the colloidal dispersion of ice particles is achieved together with tiny entrapped air bubbles in an emulsion consisting of fats, sugar, and thickening agents (polysaccharides).

Composition of Milk and Ice Cream

	Ice Cream	Milk
Fats	13	4
Nonfats	10	8
Sugar	16	—
Emulsifier	0.4	—
Stabilizer	1	—
Inorganic salts	1	1

The structure of ice cream has been studied in detail using electron microscopy. Trapped air bubbles are found to be separated by only few micrometer-thick layers of the continuous phase.

Ice cream consists of ca. 40% air-frozen foam. The continuous phase in foam consists of sugars, proteins, and emulsifiers. A typical ice cream contains

Fats	9%
Milk	10%
Sugar	15%
Emulsifiers	1%
Salts	1%

Ice cream emulsion has a very characteristic degree of stability. The air bubbles should remain dispersed, but as soon it melts in the mouth, the emulsion should break. This leads to the sensation of taste, which is very essential to enjoy its specialness. The sensation of taste on the surface of the tongue is known to be related to molecular shape and physicochemical properties. As soon as these molecules are separated from the emulsion, the taste sensation is recorded in the brain. Therefore, the various components must stay in the same phase after the breakup of the emulsion. Emulsifiers that are generally used have low HLB values (for W/O), and have been found to have considerable effect on the structure of the ice cream.

In food emulsions, one comes across a variety of products in which interfaces play an important role. One such example is churning, where air bubbles are whipped into milk or cream, the process being the same as that employed in the case of whipped cream. A large amount of air is incorporated into the liquid in bubble form, and fat globules (being surface active) collect in the bubble walls. This shows that surface-active molecules, such as fats, are collected at the air–water interface under churning or whipping. However, where whipped cream is kept cold and the agitation stopped when a stable, airy foam is produced, churned cream is warmed to the point that the globules soften and, to some degree, liquify. The ideal temperature range is said to be 12°C to 18°C. Persistent agitation knocks the softened globules into each other long enough to break through the protective membrane, and the liquid fat cements the exposed droplets together. The foam structure is broken both by the free fat and the released membrane materials, which include emulsifiers such as lecithin. These materials disrupt thin water layers and so burst bubble walls, and once enough of them have been freed in the process of whipping or churning cream, the foam will never be stable again. As churning continues, the foam gradually subsides, and the butter granules are worked together into larger and larger masses. Paddles slowly agitate the cream, causing it to thicken and separate into butter grains and buttermilk. Cold water at 10°C is then added, and it is agitated again. The added water is necessary to help the cream to "break," but the water should not exceed 25% of the total volume of the cream. Churning continues until the butter granules are about the size of wheat grains. In butter, the fat globules vary from 0.1 μm to 10 μm in diameter. The fat globule membrane comprises surface-active materials such as phospholipids and lipoproteins.

Fat globules typically aggregate in three ways:

Flocculation
Coalescence
Partial coalescence

Actually during the process of churning, one may conclude that the process of butter making can be described as an inversion of the original cream emulsion. The system of fat droplets dispersed in water is converted into a continuous phase of fat that contains water droplets. The final product is about 80% milk fat, 18% water, and 2% milk solids, mainly proteins and salts carried in the water. The physical structure of butter is found to be a bit more complicated. The continuous, amorphous phase of solid fat surrounds not only water droplets but also air bubbles, intact fat globules, and highly ordered crystals of milk fat that have grown during the cooling process. The proportion of continuous or "free" fat can vary from 50% of the total to nearly 100%, and it has a direct influence on the behavior of butter. The more fat there is in discrete globules or crystals, the harder and more crumbly the butter, even to the point of brittleness. The difference is a matter of both large-scale and molecular arrangements. In a mass where the free fat merely fills the small interstices between globules and crystals, the texture will be largely that of the separate particles. Further, it takes more energy to separate the molecules ordered in a crystal than it does to disrupt an already disordered phase of the same molecules. Mostly crystalline butter, then, will be relatively stiff and not as smooth as mostly amorphous butter. The ideal, of course, lies somewhere between the two extremes and is attained by manipulating the cooling process (much as one controls the texture of candy).

9.5.3.1   Milk—Composition and Emulsion Chemistry

The role of milk in nature is to nourish and to provide immunological protection to the mammalian young. The nutritional value of milk is high. It comes from various sources, such as human, goat, buffalo, sheep, and yak, and has been a food for humans since prehistoric times. It is also a very complex food with many thousands of different molecular species found in it. There are several factors that can affect milk composition. An approximate composition of milk can be given as follows:

Water	87%
Milkfat	4%
Solids-not-fat	9%
Protein	3%
Lactose	5%
Minerals	0.6% (Ca, P, citrate, Mg, K, Na, Zn, Cl, Fe, Cu, sulfate, bicarbonate, and other kinds)
Acids	0.1% (citrate, formate, acetate, lactate)
Enzymes	peroxidase, catalase, phosphatase, lipase
Gases	oxygen, nitrogen
Vitamins	A, C, D, thiamine, riboflavin, etc.

The following terms are used to describe milk and its fractions:

Plasma = milk − fat (skim milk)
Serum = plasma − casein micelles (whey)
Solids-not-fat (SNF) = proteins, lactose, minerals, acids, enzymes, vitamins
Total milk solids = fat + SNF

It has been found that both the composition and physical structure are important in determining the properties (colloidal or emulsion) of milk. As found in its role in nature, milk is in a liquid form (at the body temperature). We may find this observation curious if we take into consideration the fact that milk has less water than most fruits and vegetables. It is useful to analyze milk as follows:

An oil-in-water emulsion with the fat globules dispersed in the continuous serum phase
A colloid suspension of casein micelles, globular proteins, and lipoprotein particles
A solution of lactose, soluble proteins, minerals, vitamins, and other components

On image analyses of milk under a microscope at low magnification (5×), a uniform but turbid liquid is observed. At 500× magnification, spherical droplets of fat, known as fat globules, can be seen. At even higher magnification (50,000×), casein micelles can be observed. The main structural components of milk—fat globules and casein micelles—will be examined in more detail later. Milk fatty acids originate either from microbial activity in the rumen and are transported to the secretory cells via the blood and lymph, or from synthesis in the secretory cells. The main milk lipids are a class called *triglycerides,* which are composed of a glycerol backbone binding up to three different fatty acids. The fatty acids are composed of a hydrocarbon chain and a carboxyl group. The major fatty acids found in milk are

Long chain
C_{14} myristic 11%
C_{16} palmitic 26%
C_{18} stearic 10%
$C_{18:1}$ oleic 20%
Short chain (11%)
C4 butyric*
C6 caproic
C8 caprylic
C10 capric

Saturated fatty acids (no double bonds), such as myristic, palmitic, and stearic, make up two-thirds of milk fatty acids. Oleic acid is the most abundant unsaturated fatty acid in milk, with one double bond. Triglycerides account for 98% of milk fat. The small amounts of mono-, diglycerides, and free fatty acids in fresh milk may be

---

* Butyric fatty acid is specific for milk fat of ruminant animals and is responsible for the rancid flavor when it is cleaved from glycerol by lipase action.

a product of early lipolysis or simply incomplete synthesis. Other classes of lipids include phospholipids (0.8%), which are mainly associated with the fat globule membrane, and cholesterol (0.3%), which is mostly located in the fat globule core.

The physical properties of milk fat can be summarized as follows:

Density at 20°C is 915 kg/m(−3).
Refractive index (589 nm) is 1.462, which decreases with increasing temperature.
Solubility of water in fat is 0.14% (w/w) at 20°C, and increases with increasing temperature.
Thermal conductivity is about 0.17 J m(−1) s(−1) K(−1) at 20°C.
Specific heat at 40°C is about 2.1 kJ kg^{-1} K^{-1}.
Electrical conductivity is <10(exp −12) ohm(exp −1) cm(exp −1).
Dielectric constant is about 3.1.

At room temperature, the lipids are solid, and therefore, are correctly referred to as "fat" as opposed to "oil," which is in liquid form at room temperature. The melting points of individual triglycerides ranges from −75°C for tributyric glycerol to 72°C for tristearin. However, the final melting point of milk fat is at 37°C because higher-melting triglycerides dissolve in the liquid fat. This temperature is significant because 37°C is the body temperature of the cow, and the milk would need to be liquid at this temperature. The melting curves of milk fat are complicated by the diverse lipid composition:

Trans unsaturation increases melting points.
Odd-numbered and branched chains decrease melting points.
Crystallization of milk fat largely determines the physical stability of the fat globule and the consistency of high-fat dairy products, but crystal behavior is also complicated by the wide range of different triglycerides.

9.5.3.2 Milk Fat Structure—Fat Globules

More than 95% of the total milk lipid is in the form of a globule ranging in size from 0.1 to 15 μm in diameter. These liquid fat droplets are covered by a thin membrane, 8 to 10 nm in thickness, whose properties are completely different from those of both milk fat and plasma. The native fat globule membrane (FGM) is composed of apical plasma membrane of the secretory cell, which continually envelops the lipid droplets as they pass into the lumen. The major components of the native FGM, therefore, are protein and phospholipids. The phospholipids are involved in the oxidation of milk. There may be some rearrangement of the membrane after release into the lumen, as amphiphilic substances from the plasma adsorb onto the fat globule and parts of the membrane dissolve into either the globule core or the serum. The FGM decreases the lipid–serum interface to very low values, 1 to 2.5 mN/m, preventing the globules from immediate flocculation and coalescence, as well as protecting them from enzymatic action. It is well known that, if raw milk or cream is left to stand, it will separate. Stokes' Law predicts that fat globules will cream due to the differences in densities between the fat and plasma phases of milk. However, in cold raw milk, creaming takes place faster than is predicted from this fact alone.

IgM, an immunoglobulin in milk, forms a complex with lipoproteins. This complex, known as *cryoglobulin,* precipitates onto the fat globules and causes flocculation. The process is known as cold agglutination. As fat globules cluster, the speed of rising increases and sweeps up the smaller globules with them. The cream layer forms very rapidly, within 20 to 30 min, in cold milk.

Homogenization of milk prevents this creaming by decreasing the diameter and size distribution of the fat globules, causing the speed of rise to be similar for the majority of the globules. Also, homogenization causes the formation of a recombined membrane that is much similar in density to the continuous phase. Recombined membranes are very different from native FGM. Processing steps such as homogenization decreases the average diameter of the fat globule and significantly increases the surface area. Some of the native FGM will remain adsorbed, but there is no longer enough of it to cover all of the newly created surface area. Immediately after disruption of the fat globule, the surface tension raises to a high level of 15 mN/m, and amphiphilic molecules in the plasma quickly adsorb to the lipid droplet to lower this value. The adsorbed layers consist mainly of serum proteins and casein micelles.

While homogenization is the principal method for achieving stabilization of the fat emulsion in milk, fat destabilization is necessary for structure formation in butter, whipped cream, and ice cream. Fat destabilization refers to the process of clustering and clumping (partial coalescence) of the fat globules, which leads to the development of a continuous internal fat network or matrix structure in the product. Fat destabilization (sometimes "fat agglomeration") is a general term that describes the summation of several different phenomena. These include the following: *coalescence* is an irreversible increase in the size of fat globules and a loss of identity of the coalescing globules. *Flocculation* is a reversible (with minor energy input) agglomeration/clustering of fat globules with no loss of identity of the globules in the flocculation; the fat globules that flocculate can be easily redispersed if they are held together by weak forces, or they might be harder to redisperse if they share part of their interfacial layers. Partial coalescence is an irreversible agglomeration/clustering of fat globules held together by a combination of fat crystals and liquid fat, and a retention of identity of individual globules as long as the crystal structure is maintained (i.e., temperature dependent; once the crystals melt, the cluster coalesces). They usually come together in a shear field, as in whipping, and it is envisioned that the crystals at the surface of the droplets are responsible for causing colliding globules to stick together while the liquid fat partially flows between them and acts as the "cement." Partial coalescence dominates structure formation in whipped, aerated dairy emulsions, and it should be emphasized that crystals within the emulsion droplets are responsible for its occurrence.

9.5.3.3 Milk Lipids—Functional Properties

Like all fats, milk fat provides lubrication. They impart a creamy-mouth feel as opposed to a dry texture. Fat globules produce a "shortening" effect in cheese by keeping the protein matrix extended to give a soft texture. Milk proteins are one of the most important constituents. The primary structure of proteins consists of a polypeptide chain of amino acids residues joined together by peptide linkages, which may also be crosslinked by disulfide bridges. Amino acids contain both a weakly basic amino group, and a weakly acid carboxyl group both connected to a hydrocarbon chain, which is unique to different amino acids. The three-dimensional organization of proteins, or conformation, also involves secondary, tertiary, and quaternary structures. The secondary structure refers to the spatial arrangement of amino acid residues that are near one another in the linear sequence. The alpha-helix and β-pleated sheet are examples of secondary structures arising from regular and periodic steric relationships. The tertiary structure refers to the spatial arrangement of amino acid residues that are far apart in the linear sequence, producing further coiling and folding. If the protein is tightly coiled and folded into a somewhat spherical shape, it is called a globular protein. If the protein consists of long polypeptide chains that are intermolecularly linked, they are called fibrous proteins. There are some minor proteins that are associated with the FGM.

The concentration of proteins in milk is as follows:

	Grams/Liter	% of Total Protein
Total protein	33	100
Total caseins	26	80
alpha s1	10	31
alpha s2	3	8.0
beta	9	28
kappa	3	10
Total whey proteins	6	19
alpha lactalbumin	1	4
beta lactoglobulin	3	10
BSA	0.4	1.2
Immunoglobulins	1	2.1
Proteose peptone	1	2.4

Caseins, as well as their structural form—casein micelles, whey proteins, and milk enzymes are very important colloidal molecules in milk. The casein content of milk represents about 80% of milk proteins. Caseins exhibit solubility at pH 4.6. The common compositional factor is that caseins are conjugated proteins, most with phosphate group(s) esterified to serine residues. These phosphate groups are important to the structure of the casein micelle. Calcium binding by the individual caseins is proportional to the phosphate content. The conformation of caseins is much like that of denatured globular proteins. The high number of proline residues in caseins causes particular bending of the protein chain and inhibits the formation of close-packed, ordered secondary structures. Caseins contain no disulfide bonds. As well, the lack of tertiary structure accounts for the stability of caseins against heat denaturation because there is very little structure to unfold. Without a tertiary structure there is considerable exposure of hydrophobic residues. This results in strong association reactions of the caseins and renders them insoluble in water. Accordingly, strong amphiphilic protein acts like a detergent molecule. Self-association is found to be temperature dependent; and will form a large polymer at 20°C but not at 4°C. Less sensitive to calcium precipitation.

9.5.3.4 The Casein Micelle

It was mentioned earlier that some specific molecules, such as soaps or detergents, aggregate to form micelles. Most, but not all, of the casein proteins exist in a colloidal particle known as the *casein micelle*. Its biological function is to carry large amounts of highly insoluble CaP to mammalian young in liquid form and to form a clot in the stomach for more efficient nutrition. Besides casein protein, calcium, and phosphate, the micelle also contains citrate, minor ions, lipase and plasmin enzymes, and entrapped milk serum. These micelles are rather porous structures, occupying about 4 mL/g and 6–12% of the total volume fraction of milk. The "casein submicelle" model has been prominent for the past several years and is illustrated and described with the following link, but there is no universal acceptance of this model and mounting research evidence to suggest that there is no defined submicellar structure to the micelle at all. In the submicelle model, it is thought that there are small aggregates of whole casein, containing 10 to 100 casein molecules called *submicelles*. It is thought that there are two different kinds of submicelles: with and without kappa-casein. These submicelles contain a hydrophobic core and are covered by a hydrophilic coat that is at least partly composed of the polar moieties of kappa-casein. The hydrophilic CMP of the kappa-casein exists as a flexible hair. The open model also suggests there are more dense and less dense regions within the micelle, but there is less of a well-defined structure. In this model, calcium phosphate nanoclusters bind caseins and provide for the differences in density within the casein micelle.

Colloidal calcium phosphate (CCP) acts as a cement between the hundreds or even thousands of submicelles that form the casein micelle. Binding may be covalent or electrostatic. The casein micelles are not static; there are three dynamic equilibria between the micelle and its surroundings:

The free casein molecules and submicelles
The free submicelles and micelles
The dissolved colloidal calcium and phosphate

The following factors must be considered when assessing the stability of the casein micelle: The role of Ca^{++} is very significant in milk. More than 90% of the calcium content of skim milk is associated in some way or another with the casein micelle. The removal of Ca^{++} leads to reversible dissociation of β-casein without micellular disintegration. The addition of Ca^{++} leads to aggregation. The same reaction occurs between the individual caseins in the micelle, but not as much because there is no secondary structure in casein proteins.

No cysteine residues are found for alpha(s1) and β-caseins do. If any S-S bonds occur within the micelle, they are not the driving force for stabilization. Caseins are among the most hydrophobic proteins, and there is some evidence to suggest that they play a role in the stability of the micelle. It must be remembered that hydrophobic interactions are very temperature sensitive.

Electrostatic interactions: Some of the subunit interactions may be the result of ionic bonding, but the overall micellar structure is very loose and open.

Van der Waals forces: There has been some success in relating these forces to micellar stability. However, the steric stabilization has been found to be also of some importance. Especially, the hairy layer interferes with the interparticle approach. There are several factors that will affect the stability of the casein micelle system:

Salt content: It affects the calcium activity in the serum and calcium phosphate content of the micelles.

pH: Lowering the pH leads to dissolution of calcium phosphate until, at the iso-electric point (pH 4.6), all phosphate is dissolved and the caseins precipitate.

Temperature: At 4°C, the beta-casein begins to dissociate from the micelle; at 0°C, there is no micellar aggregation; freezing produces a precipitate called cryocasein.

The heat treatment leads to whey proteins becoming adsorbed, altering the behavior of the micelle. Dehydration by ethanol, for example, leads to aggregation of the micelles.

9.5.3.4.1 Casein Micelle Aggregation

Caseins are able to aggregate if the surface of the micelle is reactive. Although the casein micelle is fairly stable, there are four major ways in which aggregation can be induced:

Proteolytic enzymes as in cheese manufacturing
Acid
Heat
Age gelation

During the secondary stage, the micelles aggregate. This is due to the loss of steric repulsion of the kappa-casein as well as the loss of electrostatic repulsion due to the decrease in pH. As the pH approaches its isoelectric point (pH 4.6), the caseins aggregate. The casein micelles also have a strong tendency to aggregate because of hydrophobic interactions. Calcium assists coagulation by creating isoelectric conditions and by acting as a bridge between micelles. The temperature at the time of coagulation is very important to both the primary and secondary stages. With an increase in temperature up to 40°C, the rate of the rennet reaction increases. During the secondary stage, increased temperatures increase the hydrophobic reaction. The tertiary stage of coagulation involves the rearrangement of micelles after a gel has formed. There is a loss of paracasein identity as the milk curd firms and syneresis begins. It has been found that acidification causes the casein micelles to destabilize or aggregate by decreasing their electric charge to that of the isoelectric point. At the same time, the acidity of the medium increases the solubility of minerals so that organic calcium and phosphorus contained in the micelle gradually become soluble in the aqueous phase. Casein micelles disintegrate, and casein precipitates. Aggregation occurs as a result of entropically driven hydrophobic interactions. At temperatures above the boiling point, casein micelles will irreversibly aggregate. On heating, the buffer capacity of milk salts change, carbon dioxide is released, organic acids are produced, and tricalcium phosphate and casein phosphate may be precipitated with the release of hydrogen ions.

Age gelation is an aggregation phenomenon that affects shelf-stable, sterilized dairy products, such as concentrated milk and UHT milk products. After weeks to months of storage of these products, there is a sudden sharp increase in viscosity, accompanied by visible gelation and irreversible aggregation of the micelles into long chains forming a three-dimensional network. The actual cause and mechanism is not yet clear; however, some theories exist.

Proteolytic breakdown of the casein: Bacterial or native plasmin enzymes that are resistant to heat treatment may lead to the formation of a gel.

Chemical reactions: Polymerization of casein and whey **proteins are due to some kind of chemical reactions**. The different proteins as found in the supernatant of milk after precipitation at pH 4.6 are collectively called whey proteins. These globular proteins are more water soluble than caseins and are subject to heat denaturation. Denaturation increases their water-binding capacity. The principal fractions are β-lactoglobulin, α-lactalbumin, bovine serum albumin (BSA), and immunoglobulins (Ig).

β-*Lactoglobulins:* (MW—18,000; 162 residues) This group, including eight genetic variants, comprises approximately half the total whey proteins. β-Lactoglobulin has two internal disulfide bonds and one free thiol group. Enzymes are a group of proteins that have the ability to catalyze chemical reactions and the speed of such reactions. The action of enzymes is very specific. Vitamins in milk emulsion are one of the most important ingredients. Vitamins are organic substances essential to many life processes. Milk includes fat-soluble vitamins A, D, E, and K. Vitamin A is derived from retinol and β-carotene. Because milk is an important source of dietary vitamin A, fat-reduced products that have lost Vitamin A with the fat are required to supplement the product with Vitamin A. Milk is also an important source of the following dietary water-soluble vitamins:

B1—thiamine
B2—riboflavin
B6—pyridoxine
B12—cyanocobalamin, niacin, pantothenic acid

The major vitamin content of fresh milk is as follows:

Vitamin	Contents per Liter
A (μg RE)	400
D (IU)	40
E (μg)	1000
K (μg)	50
B1 (μg)	450
B2 (μg)	1750
Niacin (μg)	900
B6 (μg)	500

Vitamin	Contents per Liter
Pantothenic acid (μg)	3500
Biotin (μg)	35
Folic acid (μg)	55
B12 (μg)	4.5
C (mg)	20

The most remarkable observation is that all 22 minerals considered to be essential to the human diet are present in milk. Some of these are sodium (Na), potassium (K), and chloride (Cl). The electroneutrality of milk is maintained by free ions (negatively charged to lactose).

The viscosity of milk and milk products is reported to be important in the rate of creaming. The viscosity of milk increases with decrease in temperature because the increased voluminosity of casein micelles' temperatures above 65°C increases viscosity due to the denaturation of whey proteins' pH; an increase or decrease in the pH of milk also causes an increase in casein micelle voluminosity. Fat globules that have undergone cold agglutination may be dispersed due to agitation, causing a decrease in viscosity.

Optical properties provide the basis for many rapid, indirect methods of analysis such as approximate analysis by infrared absorbency or light scattering. Optical properties also determine the appearance of milk and milk products. Light scattering by fat globules and casein micelles causes milk to appear turbid and opaque. Light scattering occurs when the wavelength of light is near the same magnitude as the particle. Thus, smaller particles scatter light of shorter wavelengths. Skim milk appears slightly blue because casein micelles scatter the shorter wavelengths of visible light (blue) more than the red. The carotenoid precursor of vitamin A, β-carotene, contained in milk fat, is responsible for the "creamy" color of milk. Riboflavin imparts a greenish color to whey.

9.5.3.5 Colloids in the Food Industry

In the processing of foods and additives, rheological and mechanical properties, which determine end use properties are of primary significance. These properties are also related to others such as the impact of taste. Taste is the perception on the taste buds on the *surface* of the tongue, and colloidal properties will thus have an important role. Gels are also known to be used in many food products. The stability of food products is determined by colloidal stability in many systems (Dickinson, 1992; Friberg, 1976).

9.5.3.6 Emulsion Stability

The degree of stability of any emulsion is related to the rate of coagulation of two drops (O/W: oil drops; W/O: water drops). This process means that two oil drops in an O/W emulsion come close together, and if the repulsion forces are smaller than the attraction forces, only then the two particles meet and fuse into one larger drop. In the case of charged drops, an electrical double layer (EDL) will be present around these drops. A negatively charged oil drop (the charge arising from the emulsifier)

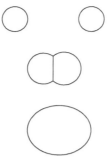

FIGURE 9.6   Stable emulsion formation by polymers. Polymer molecules adsorb (on solid particles) and penetrate (on liquid drops).

will strongly attract positively charged ions in the surrounding bulk aqueous phase. At a close distance from the surface of a drop, the distribution of the charges will be changing very much. While at a very large distance there will be electrical neutrality, there will be an equal number of positive and negative charges.

Electrostatic repulsion exists between the two negatively charged drops, and strong repulsion would be exhibited even at large distances (many times the size of the particle). The shape of the EDL curve will be dependent on the negative and positive charge distribution. It is easily seen that, if the concentration of counterions increases, the magnitude of EDL will decrease, and this will reduce the maximum of the total potential curve. The stability of emulsions can thus be increased by decreasing the counterion concentration.

Another important emulsion stabilization is achieved by using polymers. The large polymer molecules adsorbed on solid particles (Figure 9.6) will exhibit repulsion at the surface of the particles. The charged polymers also will give additional charge–charge repulsion. Polymers are used in many pharmaceutical, cosmetics, and other systems (milk). Obviously, the choice of a suitable polymer is specific to each system.

9.5.3.6.1   Advanced Emulsion Technology

As is already obvious, emulsion technology is a state-of-art system and mostly protected by patents (Johnson, 1979; Sjoblom et al., 2008; Friberg, 1976). In the following text, some examples are given that can serve as useful basic information.

9.5.3.6.1.1   Nanoemulsion Technology   The low-energy emulsification method (Emulsion Inversion Point [EIP]) has been used to prepare O/W *nanoemulsions* in the water/potassium oleate–oleic acid–C12E10/hexadecane ionic system. This method had not been used practically in ionic systems up until now. The resulting droplet sizes, much smaller than those obtained with high-energy emulsification methods, depend on the composition (formulation variables) and preparation variables (addition and mixing rate). Phase diagrams, rheology measurements, and experimental designs applied to nanoemulsion droplet sizes obtained were combined to study the formation of these nanoemulsions. To obtain nanosize droplets, it was necessary to cross a direct cubic LC phase along the emulsification path, and it is also crucial to remain in this phase long enough to incorporate all of the oil into the liquid crystal. When the nanoemulsion forms, the oil is already intimately mixed with all of the components and only has to be redistributed. Results show that smaller droplet sizes are obtained when the LC zone is wide and extends to have high water content because, in this case, during the emulsification process, the system remains long enough in the LC phase to allow the incorporation of all of the oil. Around the optimal formulation variables, the LC zone crossed by the system during emulsification is wide enough to incorporate all of the oil, whatever mixing or stirring rate is used, and then the resulting droplet size is independent of the preparation variables. However, when the composition is far from this optimum, the LC zone becomes narrower, and the mixing of components controls nanoemulsion formation. High agitation rates and/or low addition rates are required to ensure the dissolution of all of the oil into this phase.

9.5.3.6.1.2 Microemulsion Technology for Oil Reservoirs A new microemulsion additive has been developed that is effective in remediating damaged wells and that is highly effective in fluid recovery and relative permeability enhancement when applied in drilling and stimulation treatments at dilute concentrations. The microemulsion is a unique blend of biodegradable solvent, surfactant, cosolvent, and water. The nanometer-sized structures are modeled with structures that, when dispersed in the base treating fluid of water or oil, permit a greater ease of entry into a damaged area of the reservoir or fracture system. The structures maximize surface energy interaction by expanding to 12 times their individual surface areas to allow maximum contact efficiency at low concentrations (0.1–0.5%). Higher loadings on the order of 2% can be applied in the removal of water blocks and polymer damage. Laboratory data are available for the microemulsion in speeding the cleanup of injected fluids in tight gas cores. Further tests show that the microemulsion additive results in lower pressures to displace fracturing fluids from propped fractures, resulting in lower damage and higher production rates. This reduced pressure is also evident in pumping operations in which friction is lowered by 10–15% when the microemulsion is added to fracturing fluids. Field examples are shown for the remediation and fracture treatment of coals, shales, and sandstone reservoirs, where productivity is increased by 20–50%, depending on the treatment parameters. Drilling examples exist in horizontal drilling where wells clean up without the aid of workover rigs, whereas offsets typically require weeks of workover.

Terpene-based microemulsion cleaning composition has been reported in some industrial applications. Oil-in-water microemulsion cleaning compositions comprising four principal components were described based on four components. These were

1. Terpene **solvent** (e.g., D-limonene)
2. An aliphatic glycol monoether **cosolvent** (e.g., dipropylene glycol monomethyl ether)
3. A mixture of nonionic **surfactants**, consisting of a capped alkylphenol ethoxylate or an ethoxylated higher aliphatic alcohol
4. A fatty acid alkanolamide and water

The cleaning composition may be used in concentrated or diluted form for cleaning soil from glass and metal parts, among others. This microemulsion shows that, by combining water and oil, metal surfaces can be cleaned effectively. This becomes possible because both oil- and water-soluble dirt is removed by the microemulsion.

10 Diverse Applications of Surface and Colloid Chemistry in Science and Industry

10.1 INTRODUCTION

In this chapter, some important systems that are complex, as compared to more fundamental examples described previously, will be explained. This is useful since scientific developments change as more and more theoretical knowledge is obtained. The examples discussed in this chapter are designed to indicate how surface and colloid chemistry have helped in the development of new applications in recent years.

It is evident that there will appear many more new areas where the application of surface and chemistry principles will be the determining factor in development. Some examples of new research are given in this chapter as an illustration of the depth of this subject.

For example, the capillary forces mentioned in Chapter 1 become extensively involved in the movement of water through a sponge. Sponges consist of many interconnected capillaries. An oil reservoir can be considered a simplified model of a sponge. If the reservoir is finely pored and sponge-like, then oil recovery is very poor (less than 30%), while if the pores are of large diameter, then recovery will be very high (over 60%).

Additionally, there are also examples where surface and colloid chemical principles are implemented in complex systems. This area of applications is expanding as a more fundamental understanding of the surface and colloid chemistry becomes evident.

Nanotechnology and surface science: In recent decades, for example, the application of nanoscience and nanotechnology has developed since the molecular studies of surfaces at nanoscale have reached a much higher level. Nanotechnology is now a popular topic in materials science, and other areas are also emerging. It has deserved this popularity because of its interdisciplinary character, calling for expertise in physics, chemistry, biology, and medicine. Many materials properties are studied in detail.

The common factor in nanotechnology is the lateral dimension, being in the nanometer (10^{-9} = m, that is 1 billionth of a meter or 1/1000th of the thickness of a paper sheet!) range of the structures studied. Atomic or molecular distances, sizes of structures in semiconductor devices, and grain sizes in nanopowders are just a few examples out of many that can be considered as nanosized dimensions.

In nanotechnology, knowledge of the structure and composition of the materials studied is a key requirement for understanding the materials properties of a small size.

10.2 APPLICATIONS OF SCANNING PROBE MICROSCOPES (STM, AFM, FFM) TO SURFACE AND COLLOIDAL CHEMISTRY

During the end of the 20th century, a surge in the development of significantly advanced techniques has advanced *nanoscience* and technology in the development of self-assembly structures—micelles, monolayers, vesicles—biomolecules, biosensors, and surface and colloidal chemistry. In fact, the current literature indicates that there is no end to this trend regarding the vast expansion in the sensitivity and level of information.

Typically, all humans feel "*seeing is believing*," so the microscope has attracted much interest for many decades in revealing what is otherwise out of sight. Its invention, and that of other visual probes, was basically initiated on the principles laid out by the telescope (as invented by Galileo) and the light-optical microscope (as invented by Hooke). Over the years, the magnification and resolution of microscopes have improved. However, for humans to understand nature at its core, there is the dimension of atoms or molecules to be investigated. This goal has been achieved, and the subject as described here will discuss developments invented only a few decades ago.

The ultimate aim of scientists has always been to be able to see molecules while active. In order to achieve this goal, the microscope should be able to operate under ambient conditions. Further, all kinds of molecular interactions between a solid and its environment (gas or liquid or solid), initially, can take place only via the surface molecules of the interface. It is obvious that, when a solid or liquid interacts with another phase, knowledge of the molecular structures at these interfaces is of interest. The term *surface* is generally used in the context of gas–liquid or gas–solid phase boundaries, while the term *interface* is used for liquid–liquid or liquid–solid phases. Furthermore, many fundamental properties of surfaces are characterized by morphology scales of the order of 1 to 20 nm (1 nm = 10^{-9} m = 10 Å (Angstrom = 10^{-8} cm).

Generally, the basic issues that should be addressed for these different interfaces are as follows:

What do the molecules of a solid surface look like, and how are the characteristics of these different from the bulk molecules? In the case of crystals, one asks about the kinks and dislocations.

Adsorption on solid surfaces requires the same information about the structure of the adsorbates and the adsorption site and configurations.

Solid-adsorbate interaction energy is also required, as is known from the Hamaker theory.

Molecular recognition in biological systems (active sites on the surfaces of macromolecule, antibody–antigen) and biological sensors (enzyme activity, biosensors).

Self-assembly structures at interfaces.
Semiconductors.

Depending on the sensitivity and experimental conditions, the methods of molecular microscopy are many and varied. The applications of these microscopes are also extensive, as with crystal structures and the three-dimensional configurations of macromolecules.

The greatest application of microscopy is in the case of surfaces and the study of molecules at the surfaces. Generally, the study of surfaces is dependent on understanding not only the reactivity of the surface but also the underlying structures that determine reactivity. Understanding the effects of different morphologies may lead to a process for the enhancement of a given morphology, and hence, to improved reaction selectivities and product yields.

Atoms or molecules at the surface of a solid have fewer neighbors as compared with atoms in the bulk phase, which is analogous to the liquid surface; therefore, the surface atoms are characterized by an unsaturated, bond-forming capability and, accordingly, are quite reactive. Until a decade ago, electron microscopy and other similarly sensitive methods provided information about the interfaces. There were always some limitations inherent in all these techniques, which needed improvement.

A few decades ago, the best electron microscope images of globular proteins were virtually little more than shapeless blobs. However, these days, due to relentless technical advances, electron crystallography is capable of producing images at resolutions close to those attained by x-ray crystallography or multidimensional nuclear magnetic resonance (NMR). In order to improve upon some of the limitations of the electron microscope, newer methods were needed. A decade ago, a new procedure for molecular microscopy was invented and will be delineated herein. The new scanning probe microscopes not only provide new kind of information than hitherto known from x-ray diffraction, for example, but these also open up a new area of research (such as in nanoscience and nanotechnology).

The basic method of these scanning probe microscopes (SPMs; Birdi, 2002a) was essentially to be able to move a tip over the substrate surface with a sensor (probe) with molecular sensitivity (nanometer) in both the longitudinal and height direction (Figure 10.1). This may be compared with the act of sensing with a finger over a surface, or more akin to the old-fashioned record player with a metallic needle (a probe for converting mechanical vibrations to music sound) on a vinyl record.

Scanning probe microscopy was invented by Binnig and Rohrer (Nobel Prize, 1986) (Birdi, 2002a). Scanning tunneling microscope (STM) was based on scanning a probe (metallic tip) since it is a sharp tip just above the substrate, while monitoring

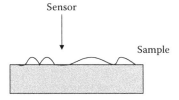

FIGURE 10.1   Basic principle of a scanning probe microscope (SPM).

some interaction between the probe and the surface. The tip is controlled to within 0.1 Å (1 nm).

In SPM, the various interactions between the tip and the substrate are as follows:

STM: The tunneling current between a metallic tip and a conducting substrate that are in very close proximity but not actually in physical contact. This is controlled by piezomotors in a stepwise method.

AFM (Atomic force microscopy): The tip is brought closer to the substrate while the van der Waals force is monitored. At a given force, the piezomotor controls this setting while the surface is scanned in the x-y direction.

FFM (Freeze Fracture Microscopy): This is a modification of AFM, where force is measured (Birdi, 2002a).

A schematic description of SPM with the tip (dimension: 0.2 mm) and sample is shown in Figure 10.2.

The most significant difference between SPM and x-ray diffraction studies has been that the former can be carried out both in the air and water (or any other fluid).

Corrosion and similar systems have been investigated using STM. The tip is covered by a plastic material, and this allows one to operate STM under a fluid environment.

STM has been used to study the molecules adsorbed on solid surfaces. The Langmuir–Blodgett (LB) films have been extensively investigated by both STM and AFM.

AFM has been used to study surface molecules under different conditions. Colloidal system studies by AFM: AFM has allowed scientists to be able to study molecular forces between molecules at very small (almost molecular size) distances. Further, it is a very attractive and sensitive tool for such measurements. In a recent study, the colloidal force as a function of pH of SiO_2 immersed in the aqueous phase was reported using AFM. The force between an SiO_2 sphere (ca. 5 mm diameter) and a chromium oxide surface in the aqueous phase of sodium phosphate were measured (pH from 3 to 11). The SiO_2 sphere was attached to the AFM sensor as shown in Figure 10.3.

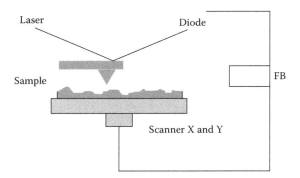

FIGURE 10.2 A schematic drawing of the sensor (tip/cantilever/optical/magnetic device) movement over a substrate in x/y/z direction with nanometer sensitivity controlled by piezomotor at the solid–gas or solid–liquid interface.

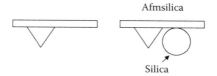

FIGURE 10.3 AFM sensor with SiO_2 sphere (schematic).

These data showed that the isolectric point (IEP) of SiO_2 was around pH 2, as expected. The binding of phosphate ions to the chrome surface were also estimated as a function of pH and ionic strength (Choi et al., 2004). Further, both STM and AFM have been used to investigate the corrosion mechanisms of metals exposed to the aqueous phase. Since both STM and AFM can operate under water, this provides a variety of possibilities.

10.2.1 Domain Patterns in Monomolecular Film Assemblies

Self-assembly monomolecular films (SAMs) at the air–water interface are structures that are very sensitive to external forces. For example, if these monomolecular assemblies are compressed to certain values, some drastic rearrangements will have to take place. Monolayers of lipids exhibit changes in structures near or after the collapse state, which has been designated as *domains*. The spontaneous formation of domain assemblies in monomolecular films of amphiphiles at the air–water (oil–water interface needs) interface has evoked great interest. During the past decade, an extensive number of investigations have been carried out by using different microscopes with varying sensitivities. The latter leads to different kinds of domain dimensional sizes.

There are two kinds of domains:

Macro domains
Molecular domains

as determined by the detection method.

Macro domains of micrometer size have been reported when using an ordinary microscope. The typical procedure is to use 2% of a fluorescent lipid analog, which renders visible a domain pattern. This, of course, assumes that the fluorescence moiety has no effect on the assembly structure. There is (generally) no information about the thickness of the domains. The shapes of domains are varied, and very complex (circular or near-circular domains, parallel stripes, or more irregular, wormlike, or similar kinds of structures).

Molecular domains have been observed using the AFM analyses of LB films of a collapsed state (Birdi, 1997, 2003). For example, cholesterol films showed *half-butterfly*-shaped domains (each domain consisting of 10^7 molecules; Figure 10.4). This quantity was estimated from the following data: the height of the domains was 90 Å, which corresponds with six layers of the cholesterol molecule (length of the molecule is found to be 15 Å from molecular models). AFM image analysis is capable

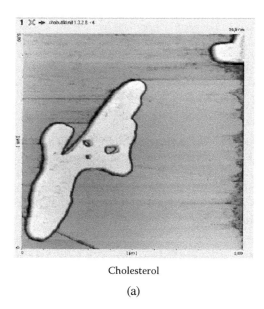

Cholesterol

(a)

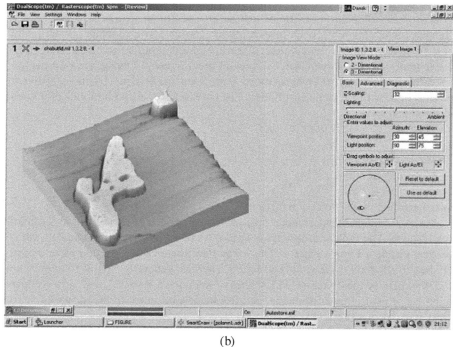

(b)

FIGURE 10.4 Cholesterol domain (consisting of 10^7 molecules) as studied by AFM. (a) Image 5 m × 5 m; (b) three-dimensional image (see text for details). The size of image is 60000 Å × 60000 Å × 100 Å.

of calculating the area of the image. Since the magnitude of the area/molecule of the cholesterol molecule is known (40 Å^2), this analysis can be carried out. In comparison to macro domains, molecular domains are measured as three-dimensional by AFM. This indicates that such nanostructures can be investigated by using AFM. Thus, in the future, any nanosurface analysis will be useful in understanding these new technologies.

As compared to ordinary microscopes, SPM provides two- and three-dimensional images. The 3D images allow one to see molecules with different diameters (Birdi, 2002a).

10.3   DRUG DELIVERY DESIGN (DDD)

In pharmaceutics, a medicine is prescribed in a quantity that can be present in the blood at a particular and sufficient concentration. If one considers some examples, the range of concentration per doses (i.e., per tablet) may range from microgram to 1000 mg. Since many of medicinal molecules decompose in the body before reaching the target (such as the liver, heart, or lungs), one uses a higher concentration (10 to 1000 times) than needed at the target. During the past decades, investigations have been reported on designing drug delivery techniques that can reduce the initial excess concentration.

10.4   SURFACE AND COLLOIDAL ASPECTS OF CEMENT INDUSTRY

Mankind has used some form of cement for thousands of years to build houses and monuments. Some examples are the pyramids (3000 BC) and the Colosseum (2000 BC), indicating an early awareness of the role of surface and colloidal chemistry. In modern times, besides small structures such as houses, the use of cement has been widened to include very large constructions such as dams. The main aim of using cement is to bind two bricks with a material consisting of very fine particles, and which hardens after water has evaporated.

In general, there are two main classes of constructional cements: nonhydraulic, which do not set under water, and hydraulic. Their composition can be indicated by the following two examples:

(i) Plaster of Paris ($CaSO_4 \cdot 1/2\ H_2O$)

$$CaSO_4 \cdot 1/2\ H_2O + 1^{1}/2\ H_2O \Rightarrow CaSO_4 \cdot 2H_2O \text{ (gypsum)}$$

(ii) Lime-based cement (CaO)

$$CaO + H_2O \Rightarrow Ca(OH)_2 + CO_2 \rightarrow CaCO_3 \text{ (calcite)}$$

Nonhydraulic cements were among the most common of the ancient cements. The relatively high solubilities of portlandite ($Ca[OH]_2$) and gypsum means that they deteriorate rapidly in moist or wet environments. Many decades ago, the Romans used lime-based cements and mortars (cement plus sand) by ramming the wet pastes to form a high-density surface layer that carbonates in contact with air to produce a low-permeability surface of calcite. The use of this procedure leads to the protection of the $Ca(OH)_2$ layer (such as seen in the Roman structure Hadrian's Wall). Lime mortars were still used in domestic construction until relatively recently. Partial dehydration of natural gypsum (200°C), and calcination of calcite (850°C).

It is likely that, during the early Roman period, silica and alumina, as well as volcanic earths, were used as cements. Portland cement is made from finely ground limestone and finely divided clay to give a burned product containing 70% CaO, 20% SiO_2, 4% Fe_2O_3, and 4% Al_2O_3, plus smaller amounts of minor oxides (e.g., Na_2O, K_2O, MgO, etc.).

The major raw materials used for modern cement making are

Limestones, argillaceous shales
Chalks, schists
Shells, clays
Calcerous muds, other iron-bearing aluminosilicates

The essential step is the efficient *grinding and blending* of raw materials. The final properties of cement strongly depend on its mineral composition so that raw composition and firing conditions are adjusted, depending on the type of cement to be produced. The microstructure of the steel fiber–cement paste interface was studied by scanning electron microscopy (SEM). The interfacial zone surrounding the fiber was found to be substantially different from the bulk paste further away from the fiber surface. The interfacial zone consisted of

1. A thin (1 or 2 m thick) duplex film in actual contact with the reinforcement.
2. Outside this, a zone of perhaps 10 to 30 m thickness, which, in reasonably well-hydrated systems, is largely occupied by relatively massive calcium hydroxide crystals, with occasional interruptions of more porous regions.
3. Outside this, a highly porous layer parallel to the interface. The interaction of cracks initiated in the matrix with this interfacial zone was observed.

In the cement industry, the term *hydration* is used to describe a range of reactions between cement and water to produce a hardened product. A cement clinker particle is a multiphase solid having massive calcium silicate grains (50–100 μm) in a matrix of interstitial aluminate and ferrite. This is described as analogous to a distorted clay sequence, which traps regions of porosity–pore size distribution from nanometer to micrometer.

The most important characteristic of cement is its pore structure and aqueous phase; hence, the microstructure of the hardened cement paste via the pore system. It is highly alkaline (pH > 13) due to rapid and almost quantitative dissolution of Na and K salts from the cement clinker. The porosity of the paste comprises interconnected and isolated pores, the pore sizes of which are important to the strength and dimensional stability of cement products. Different types of cement are used to meet different performance criteria. Properties can be estimated from compositions and fineness (i.e., particle size and size distribution). In the past, additives have been used with cement that become part of a waste disposal process. The most common of these are pulverized fly ash (PFA), a coal combustion product; blast furnace slag (BFS) from iron making; and condensed silica fume (CSF) from the ferrosilicon industry.

The hydration characteristics of cement (rate and extent) are of great importance in the end product. Silica is known to react relatively fast in the cement system.

Hydration gives rise to effects on pore filling and the consequent enhancement of mechanical performance (low-porosity pastes are stronger than high-porosity ones). The first fast hydration step is followed by a relatively dormant period that may last 6 months or more, depending on temperature, particle size, and aqueous phase composition. In order to control the hydration step, alkyl sulfonate salts surface-active substances (SAS) are used.

The durability of cement pastes is strongly influenced by (1) internal chemistry, and (2) paste microstructure. The industrial by-product additives above all influence the development of paste microstructures. Isolated pores are completely enclosed by hydration products so that material transport into and out of the pore is limited. Connected porosity is that through which a continuous pathway between regions of the microstructure exists. Continuous or interconnected porosity often (although not always) links the interior of the paste to the outside world so that aggressive chemical species can penetrate and degrade the paste internally, affecting paste durability. The effect of blending agents identified previously on microstructure is to cause a reduction in the degree of interconnected porosity. This is especially true in the case of BFS-containing pastes. Although the overall porosity, as determined by neutron scattering, is still significant, interconnected porosity, as measured by intrusion methods (e.g., mercury intrusion porosimetry [MIP]) is low.

In addition, further oxidation and cathodic reactions lead to the production of oxides and oxyhydroxides of Fe (III), which produces a low-permeability, passive film that slows down the corrosion rate considerably. Where corrosion can continue (by depassivation), the expansion of corrosion products at the cement–steel interface and the subsequent spalling of cover concrete can occur. Many examples of this can be seen in concrete structures.

Glass-fiber reinforcement corrosion: Unlike steel reinf*orcement, glass fibers are introduced in random* orientation and throughout the paste. Typically, as filament bundles (of around 50 filaments), the fiber will be of variable length (usually up to 2 cm).

Cement has been used in modern times to build houses and very large structures, such as dams. Some specifications for the Hoover Dam in the United States Southwest are as follows:

Height	250 m
Weight	6.6 million tons
Volume of concrete used	118 million ft³
Width	250 m

The surface and colloidal chemistry of cement is indeed very complicated, and the Hoover Dam was among the leading engineering achievements of the 20th century.

It is important to note that one of the greatest obstacles the builders had to overcome was concrete and cement chemistry.

Concrete is made of cement aggregate and water mixed together to form a paste. The aggregate is usually a filler material composed of inert ingredients such as sand and rocks. When water is added, the components of cement undergo a chemical reaction known as hydration. As hydration occurs, the silicates are transformed into silicate hydrates and calcium hydroxide ($Ca[OH]_2$), and the cement slowly forms a hardened paste.

One of the products of the hydration reaction is the significant heat it produces. For example, consider the amount of heat that was produced during the building of the Hoover Dam by just one major cement component, Ca_3SiO_5, when 2 mol of Ca_3SiO_5 produce 173.6 kilojoules (kJ) of heat, assuming that the percentage of Ca_3SiO_5 in the concrete is 30% cement and water (and 70% aggregate). The wet Hoover Dam concrete had a density of 59.1 kg/ft^3 and 118,000,000 ft^3 of concrete by volume. Then, with 6.97 × 10^9 kg or 6.97 × 10^{12} g of concrete, 7.5% of this number gives *us* 5.23 × 10^{11} g of Ca_3SiO_5 for the Hoover Dam. The amount of heat generated if every gram of Ca_3SiO_5 reacts can be estimated. If we convert grams Ca_3SiO_5 to moles Ca_3SiO_5, we find that Ca_3SiO_5 hydration alone will produce 200 trillion kJ or 4.8 × 10^{13} calories of heat. This corresponds to bringing 700 million liters of water at room temperature (25°C) to a boiling (100°C) temperature. It was therefore necessary to use refrigeration in order to control the temperature. On the other hand, if left to cool by itself, it was estimated that it would take about 100 years to build the dam. Thus, holes were made in the dam with 582 miles of 3 cm cooling pipe. The massive refrigeration plant that was deployed to cool the dam had the power to produce the equivalent of 1,000 tons of ice every 24 h. The procedure took the contractors a little more than 4 years to complete concrete placement in the dam.

10.5   DIVERSE INDUSTRIAL APPLICATIONS

10.5.1   PAPER INDUSTRY (SURFACE AND COLLOIDAL ASPECTS)

The paper industry today is complex and uses highly advanced industrial innovations. Not only is the thickness of the paper for printing (such as 80 or 90 or 100 gm/cm^2) specified but also the paper gloss or paper brightness. In fact, in many quantitative analyses (such as measurement of area under a curve), the weight of the paper is a central consideration. Since the weight is very reliable, it is routinely used in analyses (such as curve areas) of the surface chemical properties of the paper. During paper manufacturing, colloid chemistry is applied to control the implications of additives (such as fillers, sizing, dyes, and strength additives). The paper acquires negative charge due to oxidation when exposed to air (mostly -COO- charges, with pKa = 5). This suggests that at pH = 5, paper pulp will have almost zero charge.

The paper industry deals with various substances, such as

Cellulose
Lignin
Dyes

Dispersants
Fillers ($CaCO_3$, TiO_2 [whitener])
Alum
Defoamers

10.5.1.1 Inks and Printers (Colloidal Chemistry)

Over thousands of years for writing, the ancient people used naturally occurring colloidal fine material from ash (mostly charcoal) dispersed in oil (olive oil). Modern inkjet printers employing color are based on much more sophisticated components. Inkjet printers have a number of nozzles that inject ink droplets on the surface of paper. Simultaneously, different colors are mixed to obtain the desired color shade (more than hundreds of thousands). In a typical printer, there may be 30,000 injections per second, and there may be more than 500 nozzles (each with a size less than a human hair ($\mu m = 10^{-6}$ m). (The ink has a shelf life of more than a year.) In this process, the surface and colloidal principles most obvious are

Contact angle (ink and the nozzle; ink and the surface of the paper)
Capillary pressure (at the nozzle)

It is thus seen that surface and colloidal principles are involved in a much complex manner in these printers.

10.5.1.2 Theory of Adhesives and Adhesion

Adhesives are used in everyday applications. Adhesives may be in liquid form or thick pastes. Their main mechanism is based on the polymerization or crosslinking of polymers, which gives rise to glue or other adhesive application. The degree of adhesion of such a process is determined by conventional technological tests.

What follows from Young's equation is that, if a liquid is removed from the surface of a solid, work will be needed, to execute this process (= Wad). This process will require the destruction of 1 cm^2 of interface γSL, and the creation of 1 cm^2 of γ S and γ L. From this we get

$$Wad = \gamma S + \gamma L - \gamma SL \tag{10.1}$$

which, on combining with Young's equation, gives

$$Wad = \gamma L (1 + Cos (\theta)) \tag{10.2}$$

This shows that, to remove a liquid from a solid surface, surface tension and contact angle θ are required. If the liquid wets the solid surfaces (water on glass $\theta = 0$),

$$Wad = 2 \gamma L \tag{10.3}$$

Wad is thus the work needed to create twice the surface tension (one on each side after separation or breakup).

Another area of much interest is the adhesion of ice to solids. This system is obviously of much interest in general everyday phenomena (tire friction on road surfaces, ice on metal surfaces, ships, etc.). Especially of interest is the adhesion of ice on ships sailing in the cold areas, and on wings of airplanes. Investigations have shown that the adhesive bonds between clean metal surface and ice are very strong. When the ice is removed by force, it breaks, leaving a thin layer of ice on the solid layer.

Further, a large attraction exists between two smooth solid surfaces if a drop of liquid (say, water and two glass plates). It is thus obvious that the peeling energy of two plates in this system will increase if a glue or similar substance is used (instead of water).

10.5.1.3   Surface Coatings of Solids

The properties of solid surfaces can be changed by coating while the rest of the material remains unchanged. Such coatings are numerous, and new applications are being continuously developed.

A recent product called Sol-Gel (solution and gelling) coating has as its principle the following:

A solid surface is coated with a metal alkyl-oxide solution.
The metal alkyl-oxide is hydrolyzed and forms a gel.
The gel forms a network on the solid surface.
The gel is exposed to high temperature, leaving behind a transparent coating.

The alkyl oxide $(Si[RO]_4)n$, is converted by a catalyst and, after dealkylation, it forms a polymer of $(Si[OH]_4)n$. On heating and dehydration, a coating of $(Si[O]_4)n$ is obtained. The thickness of the coating on aluminum (or other metal surface) can be 3 to 10 μm.

10.5.2   Emulsion Polymerization

Emulsion polymerization is one of the major examples where detergents are applied to create microreactions. For instance, to polymerize styrene (which is insoluble in water), an initiator is added to the aqueous phase. The polymer (polystyrene [PS]) is formed, and the suspension is stabilized by using suitable emulsifiers. The latex thus formed is used in various industrial applications.

Polystyrene can be prepared as follows: A mixture of styrene, detergent (Na-dodecanoate), and water is agitated ultrasonically to produce a fine emulsion. On the addition of hydrogen peroxide (initiator), PS is obtained as a polymer, which can be extracted after filtration. The polymer molecular weight is determined by various methods (such as light scattering and osmotic pressure).

10.5.2.1   Polymer Colloidal Systems

Polymers play a important role in both biological and industrial systems. Recently, there has been much interest in developing methods to apply polymer colloidals. The principle is to employ polymer colloidal beads instead of using large solid particles.

These colloidals are dispersed in aqueous media, and exhibit twofold progenies: polymers and colloidals.

10.5.2.1.1 Photographic Industry (Emulsion Films)

Photographic film production has been a very sophisticated technical innovation. In general, the photographic film consists of a sheet of plastic base (polyester or nitrocellulose [celluloid]) that is coated with a very thin emulsion containing light-sensitive Ag salts. The latter is bonded to gelatine gel. The film is exposed to an image created by the lens. The resolution of the image produced is determined by the size of the Ag particles. Most of the industry is prone to high secrecy, and not much is known about how the different producers of film achieve the end result. Black-and-white films were much simpler than color films. In color films, the emulsion consisted of three different color layers. Each was developed separately after exposure. It is obvious that the demand of making such a film is very complex. Further, the chemical reactions that take place on the surface of the film may vary from minutes (or longer) to merely 1/1000 s. This, indeed, demands much technology knowhow. Therefore, not only are their production techniques safeguarded by patents but there is also a large degree of secrecy at the production sites.

Making photographic gelatin: Photographic gelatins have a suspension of silver halide salts that are sensitive to light. When they are still warm, these gelatins are spread on a transparent plastic film to obtain a photographic film, or on a card to obtain a surface for photographic prints. As shown in the history of photography, there are many methods to produce photosensitive surfaces, and many of them do not use silver salts.

10.5.2.1.2 Diverse Applications of Foams

Even though Foam formation is a simple process (i.e., blowing air into a solution of a surface-active agent in water), its applications are far too involved, especially in those cases where foam is the main characteristic of the end product. Some useful examples are discussed in the following text.

Firefighting: Most firefighting begins with water; however, water alone, in many cases, is not efficient due to several limitations. The main approach is to contain the fire by covering the flames with a layer of foam so that their contact with air (i.e., oxygen) is prevented. In the case of the application of foams in firefighting, the rate of drainage must be considered. The foam firefighting procedure is useful since it gives slower evaporation and thus better results. The thick foam also helps in reducing the vapors (of organic fluids).

Diverse foam structure applications: In foam rubber, foamed polymers, shaving foams, milk shakes, and whipped creams, slowly draining thin liquid films (TLF) are needed. Accordingly, the rate of drainage is the most important factor in such industrial foam applications.

10.5.2.2 Medical Care and Surface Chemistry

Injury to any part of the human body leads to blood loss, and in general, healing requires fast clotting in order to reduce excess loss. Each soldier in the U.S. army carries a sponge or powder that contains clotting agents. These treatments are found to reduce loss of life much more than in early wars. The powder is made of porous mineral zeolites. It is mostly surface chemical reactions that determine their function. Currently, much research is being carried out on these treatment applications.

10.5.2.2.1   Nucleation and New Phase Formation

When any liquid is cooled below its freezing point, it would be expected to simply freeze and form a solid phase. However, this phase transition is not so simple in reality. A liquid phase, on cooling below the freezing point, will change to a solid phase, or, a gas phase, on cooling, would change over to a liquid phase at the appropriate temperature and pressure. The changing of water vapor to liquid may well be a central concern to humans as adequate fresh water sources become scarcer.

However, one finds that, in cooling a liquid below its freezing point, the liquid may not always turn into solid phase at the freezing point. In fact, in some cases, such as water, even at around $-40°C$, liquid water does not turn into a solid phase. It stays in what is called a *supercooled* state. A major phenomena is the freezing of supercooled clouds. However, if certain so-called nucleating agents are used, then the clouds would turn into liquid droplets (and form rain). The nucleation process is a surface phenomena and is observed in transitions from

Gas to liquid
Liquid to solid

10.5.2.2.2   Electrokinetic Systems

The variation of charge potential near an interface induces many unique properties in charged particles (such as colloids, emulsion drops, solid–liquid interfaces). The *electro double layer* (EDL) at the interface of a solid and liquid was described earlier. Electric fields can be used to generate bulk fluid motion (electroosmosis) and to separate charged species (molecules and particles). Most solid surfaces acquire a net charge when brought into contact with an electrolyte solution. This happens, for instance, on the surface of material such as silica and glass. This net charge attracts nearby ions of opposite charge (i.e., positive ions) and, at the same time, repels ions of like charges. The typical thickness of this region (called the Stern layer) is 1 to 20 nm (Figure 10.5).

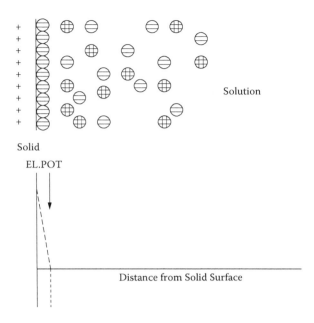

FIGURE 10.5   The state of ions in the electrical double layer (schematic).

TABLE 10.1
Areas of Application for the STM and AFM[bir97]

Lipid Monolayers (as Langmuir–Blodgett films)
Different layered substances on solids
Self-assembly structures at interfaces
Solid surfaces

Langmuir–Blodgett films
Thin-film technology
Interactions at surfaces of ion beams/laser damage
Nanoetching and lithography, nanotechnology
Semiconductors
Mineral surface morphology
Metal surfaces (roughness)
Microfabrication techniques
Optical and compact discs
Ceramic surface structures
Catalyses
Surface adsorption (metals, minerals)
Surface manipulation by STM/AFM
Polymers
Biopolymers (peptides, proteins, DNA, cells, virus)
Vaccines

JIM—Should this table be included? If so, where? (Include intext citation.) If not, just mark it for deletion in Final Pages.

References

Adam, N. K., *The Physics and Chemistry of Surfaces*, Clarendon Press, Oxford, 1930.

Adamson, A. W. and Gast, A. P., *Physical Chemistry of Surfaces*, 6th ed., Wiley-Interscience, New York, 1997.

Aveyard, R. and Hayden, D. A., *An Introduction to Principles of Surface Chemistry*, Cambridge University Press, London, 1973.

Baker, L., in *Proceed. Int. Symp., Future Electron Devices,* Tokyo, 1985.

Bakker, G., *Kapillaritat und Uberflachenspannung Handbuch der Eksperimentalphysik*, 3d vi, Leipzig, 1928; Freundlich, H., *Colloid and Capillary Chemistry*, Methuen, London, 1926.

Bancroft, W. D., *Applied Colloid Chemistry*, McGraw-Hill, New York, 1932.

Birdi, K. S., *J. Colloid Polym. Sci.*, 250, 7, 1972.

Birdi, K. S., *J. Colloid Polym. Sci.*, 260, 8, 1982.

Birdi, K. S., *Lipid and Biopolymer Monolayers at Liquid Interfaces*, Plenum Press, New York, 1989.

Birdi, K. S., *Self-Assembly Monolayer (SAM) Structures*, Plenum Press, New York, 1999.

Birdi, K. S., *Fractals in Chemistry, Geochemistry, and Biophysics*, Plenum Press, New York, 1993.

Birdi, K. S., *Scanning Probe Microscopes*, CRC Press, New York, 2002a.

Birdi, K. S., Ed., *Handbook of Surface and Colloid Chemistry*, CRC Press, Boca Raton, FL, 1997.

Birdi, K. S., Ed., *Handbook of Surface and Colloid Chemistry*—CD Rom, CRC Press, Boca Raton, FL, 2002 (Second Edition).

Birdi, K. S., Ed., *Handbook of Surface and Colloid Chemistry*—CRC Press, Boca Raton, FL, 2009 (Third Edition).

Birdi, K. S., Vu, D. T., Winter, A., and Naargard, A., *J. Colloid Polym. Sci.*, 266, 5, 1988.

Biresaw, G. and Mittal, K. L., *Surfactants in Tribology*, CRC Press, New York, 2008.

Blank, M., *J. Colloid Interface Sci.,* 30, 233, 1970.

Boys, C. V., *Soap Bubbles*, Dover Publications, New York, 1959.

Chattoraj, D. K. and Birdi, K. S., *Adsorption and the Gibbs Surface Excess*, Plenum Press, New York, 1984.

Choi, W., Lee, S. M., and Singh, R. K., *Electrochem. Solid-State Lett.,* 7, 141, 2004.

Cini, R., Loglio, G., and Ficalbi, A., *J. Colloid Interface Sci.*, 41, 287, 1972.

Davies, J. T. and Rideal, E. K., *Interfacial Phenomena*, Academic Press, New York, 1963.

Defay, R., Prigogine, I., Bellemans, A., and Everett, D. H., *Surface Tension and Adsorption*, Longmans, Green, London, 1966.

Fanum, M., *Microemulsions*, CRC Press, New York, 2008.

Faraday Symposia of the Chemical Society, No. 16, *Structure of the Interfacial Region*, Faraday Society, London, 1981.

Feder, J., *Fractals: Physics of Solids and Liquids*, Plenum Press, New York; 1988; Avnir, D., Ed., *The Fractal Approach to Heterogeneous Chemistry*, John Wiley & Sons, New York, 1989.

Fendler, J. H. and Fendler, E. J., *Catalysis in Micellar and Macromolecular Systems*, Academic Press, New York, 1975.

Friberg, S., Larsson, K., and Sjoblom, J., *Food Emulsions,* CRC Press, Boca Raton, 2003.

Gaines, G. L., Jr., *Insoluble Monolayers at Liquid–Gas Interfaces*, Wiley-Interscience, New York, 1966.

Gibbs, J. W., *Collected Works,* 2nd Ed., Vol. 1, Longmans, New York, 1928.

Gitis, N. and Sivamani, R., *Tribology Transactions,* Taylor & Francis, New York, 2004.

Guggenheim, E. A., *Trans. Farad. Soc.,* 41, 150, 1945.

Harkins, W. D., *The Physical Chemistry of Surface Films*, Reinhold, New York, 1952.

Holmberg, K., Ed., *Handbook of Applied Surface and Colloid Chemistry,* John-Wiley & Sons, Indianapolis, 2002.

Ivanov, I., *J. Dispersions Sci. Technol.,* 9, 321, 1988.

Jaycock M. K. and Parfitt, G. D., *Chemistry of Interfaces,* John Wiley & Sons, New York, 1981.

La Mer, V. K., Ed., *Retardation of Evaporation by Monolayers,* Academic Press, New York, 1962.

Lovett, D., *Science with Soap Films,* Institute of Physics Publishing, Bristol, U.K., 1994.

Matijevic, E., Ed., *Surface and Colloid Science,* Vol. 1B9, Wiley-Interscience, New York, 1969–1976.

Miller, C. A. and Neogi, P., *Interfacial Phenomena,* CRC Press, New York, 2008.

Partington, J. R., *An Advanced Treatise of Physical Chemistry,* Vol. II, Longmans, Green, New York, 1951.

Ruiz, C. C., *Sugar-Based Surfactants,* CRC Press, New York, 2008.

Sjoblom, J., in *Handbook of Surface and Colloid Chemistry* (Ed. K. S. Birdi), 3rd ed., CRC Press, Boca Raton, 2008.

Soltis, A. N., Chen, J., Atkin, L. Q., and Hendy, S., *Curr. Appl. Phys.,* 4, 152, 2004.

Somasundaran, P., *Colloidal and Surfactant Sciences,* CRC Press, New York, 2006.

Stefan, J., *Ann. Phys.,* 29, 655, 1886.

Tanford, C., *The Hydrophobic Effect,* John Wiley, New York, 1980.

Wu, X. Z., Ocko, B. M., Sirota, C. B., Sinha, S. K., Deutsch, M., Cao, B. H., and Kim, M. W., *Science,* 261, 1018, 1993.

Zana, R., *Giant Micelles,* CRC Press, New York, 2008.

Zoller, U., *Sustainable Development of Detergents,* CRC Press, New York, 2008.

Appendix A: Effect of Temperature and Pressure on Surface Tension of Liquids (Corresponding States Theory)

Both in industry and research, large data are manipulated that could be systemized. Understanding the chemistry and physics of liquid surfaces is important so as to describe interfacial forces as a function of temperature and pressure. The magnitude of surface tension, γ, decreases almost linearly with temperature (**t**) within a narrow range (Birdi, 2002, 2008; Defay et al., 1966):

$$\gamma t = ko \, (1 - k1 \, \mathbf{t}) \tag{A.1}$$

where ko is a constant. It was found that coefficient k1 is approximately equal to the rate of decrease of density (ρ), with rise in temperature:

$$\rho t = \rho o \, (1 - k1 \, \mathbf{t}) \tag{A.2}$$

Where ρo is the value of density at $t = 0°C$, and values of constant k1 were found to be different for different liquids. Further, the value of γ was related to the critical temperature (TC).

The following equation relates surface tension of a liquid to the density of liquid, ρl, and vapor, ρv (Partington, 1951; Birdi, 1989):

$$\gamma/(\rho l - \rho v \,)^4 = C \tag{A.3}$$

where the value of the constant C is nonvariable only for organic liquids, while it is not constant for liquid metals.

At the critical temperature, Tc, and critical pressure, Pc, a liquid and its vapor are identical, and the surface tension, γ, and total surface energy, as in the case of the energy of vaporization, must be zero (Birdi, 1997). At temperatures below the boiling point, which is 2/3 Tc, the total surface energy and the energy of evaporation are nearly constant. The variation in surface tension, γ, with temperature is given in Figure A.1 for different liquids.

These data clearly show that the variation of γ with temperature is a very characteristic physical property. This observation becomes even more important when it is considered that the sensitivity of γ measurements can be as high as approximately 0.001 dyn/cm(= mN/m; as described in detail in the following text). The change in γ with temperature in the case of mixtures would thus be dependent on their

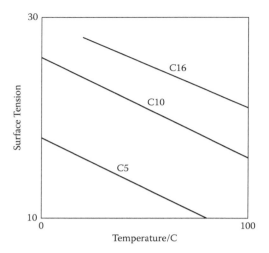

FIGURE A.1   Variation of γ with temperature for different alkanes (C_5: *n*-pentane; C_{10}: *n*-decane; C_{16}: *n*-hexadecane).

composition. The addition of gas to a liquid always decreases the value of γ. For example, the variation of γ of the system CH_4 + hexane is given as follows:

$$\gamma(CH_4 + hexane) = 0.64 + 17.85 \, x \, hexane \qquad (A.4)$$

It is seen that, actually by measuring γ for such a system, the concentration of CH_4 can be estimated. This fact has much relevance in oil reservoir engineering operations where CH_4 is found in crude oil.

It is well known that the corresponding states theory can provide much useful information about the thermodynamics and transport properties of fluids. For example, the most useful two-parameter empirical expression, which relates the surface tension, γ, to the critical temperature, Tc, is given as

$$\gamma = ko \, (1-T/Tc)^{k1} \qquad (A.5)$$

where ko and k1 are constants. Van der Waals derived this equation and showed that k1 = 3/2, although experiments indicated that k1 = 1.23. Guggenheim (Guggenheim, 1945) has suggested that k1 = 11/9 . However, for many liquids, the value of k1 lies between 6/5 and 5/4.

Van der Waals also found ko to be proportional to $Tc^{1/3} \, Pc^{2/3}$. Equation A.5, when fitted to the surface tension, γ, data of liquid CH_4, has been found to give the following relationship:

$$\gamma CH_4 = 40.52 \, (1-T/190.55)^{1.287} \qquad (A.6)$$

where Tc = 190.55 K. This equation has been found to fit the γ data for liquid methane from 91 to −190 K, with an accuracy of "0.5 mN/m. In a recent study, the γ

versus T data on n-alkanes, from n-pentane to n-hexadecane, were analyzed (Birdi, 1997). The constants ko (between 52 and 58) and k1 (between 1.2 and 1.5) were found to be dependent on the number of carbon atoms, Cn, and since Tc is also found to be dependent on Cn, the expression for all the different alkanes that individually were fit to Equation A.5 culminated in a general equation where γ was a function of Cn and T as follows (Birdi, 1997):

$$\gamma = \text{function of T, Cn} \tag{A.7}$$

$$= (41.41 + 2.731 \text{ Cn} - 0.192 \text{ Cn5} + 0.00503 \text{ Cn}^3)$$

$$X (1 - T/\{273 + (-99.86 + 145.4 \ln(\text{Cn}) + 17.05 (\ln(\text{Cn})[^5])^{k1} \tag{A.8}$$

where $k1 = 0.9968 + 0.087 \text{ Cn} - 0.00282 \text{ Cn}^5 + 0.00084 \text{ Cn}^3$. The estimated values from Equation A.8 for different n-alkanes were found to agree with the measured data within a few percent: γ for n-C_{18} H_{38}, at 100°C, was 21.6 mN/m, from both measured and calculated values. This agreement shows that the surface tension data on n-alkanes fit the corresponding state equation very satisfactorily. The physical analyses of the constants ko and k1 need to be investigated at this stage.

It is worth mentioning that the equation for the data on γ versus T, for polar (and associating) molecules like water and alcohols, when analyzed by Equation A.5, gives the magnitudes of ko and k1, which are significantly different from those found for non-polar molecules such as alkanes, etc. This difference requires further analyses so that the relationship between γ and Tc may be more completely understood (Birdi, 1997).

The variation of γ for water with temperature(t/C) is given as (Cini et al., 1972; Birdi, 1997):

$$\gamma \text{ H}_2\text{O} = 75.668 - 0.1396 \text{ t} - 0.2885 \text{ } 10^{-3} \text{ t5} \tag{A.9}$$

These data are useful since water is used as a calibration liquid in many surface tension studies.

The surface entropy (Ss) corresponding to Equation A.5 is

$$Ss = -d\gamma/dT = k1 \text{ ko } (1 - T/Tc)^{k1} - 1 /(Tc) \tag{A.10}$$

and the corresponding surface enthalpy, hs is

$$hs = gs - T \text{ Ss}$$

$$= -T (d \gamma/dT)$$

$$= ko (1 - T/Tc)^{k1-1} (1+(k1 - 1) T/Tc) \tag{A.11}$$

The reason heat is absorbed on expansion of a surface is that the molecules must be transferred from the interior against the inward attractive force to form the new surface. In this process, the motion of the molecules is retarded by this inward attraction so that the temperature of the surface layers is lower than that of the interior unless heat is supplied from outside.

Further, extrapolation of γ to zero surface tension in the data given in Figure A.1 gave values of Tc that were 10–25% lower than the measured values (Birdi, 1997, 2002). This deviation was recently analyzed in much detail (Birdi, 1997, 2002).

The surface tension, γ, of any liquid would be related to the pressure, P, as follows:

$$(d\gamma/dP)A,T = (dV/dA)P,T \tag{A.12}$$

Since the quantity on the right-hand side would be positive, the effect of pressure should be to give an increase in γ. Preliminary analyses indicates that the term (d γ/dP) is positive and dependent on the alkane chain length (Birdi, 1997). These considerations are important in many systems where high pressure is present, for example,

Oil reservoirs found at 100–200 atm pressure
High-pressure technologies
Car tires exerting high pressure on the roads
Teeth in the mouth exerting considerable high pressure
Shoes (e.g., soles) exposed to high pressures.
Building structures

The following relationship relates γ to density (Birdi, 1997):

$$\gamma (M/\rho)^{2/3} = k (Tc - T - 6) \tag{A.13}$$

where M is the molecular weight, and ρ is the density (M/ρ = molar volume). The quantity ($\gamma [M/\rho]2/3$) is called the *molecular surface energy*. It is important to notice the correction term 6 on the right-hand side. This is the same as found for *n*-alkanes and *n*-alkenes in the estimation of Tc from γ versus temperature data (Birdi, 1997).

Calculated γ [from Equation A.8] and measured values of different *n*-alkanes at various temperatures are as follows:

n-Alkane	Temperature (°C)	Measured	Calculated
C_5	0	18.23	18.25
	50	12.91	12.8
C_6	0	20.45	20.40
	60	14.31	14.3
C_7	30	19.16	19.17
	80	14.31	14.26
C_9	0	24.76	24.70
	50	19.97	20.05
	100	15.41	15.4
C_{14}	10	27.47	27.4
	100	19.66	19.60
C_{16}	50	24.90	24.90
C_{18}	30	27.50	27.50
	100	21.58	21.60

Appendix B: Solubility of Organic Molecules in Water Using a *Surface Tension–Cavity* Model System

The mechanisms of solubility in water of an inorganic salt, such as NaCl, is different from that of an alkane (such as hexane) molecule. NaCl dissociates into Na and Cl ions and interact with water through hydrogen bonds. Hexane molecules dissolve (although at a very low solubility) in water when placed inside a water structure. Since the water structure is stabilized mainly by hydrogen bonds, the hexane molecule will give rise to some rearrangements of these bonds but without breaking.

If a salt exhibits maximum solubility (called saturation solubility) of 10 mol/L in water, then it corresponds to ca. 10 mol of salt:55 mol of water (ratio of 1:5.5). On the other hand, an alkane may show a solubility of 0.0001 mol/L in water (0.0001 mol:55 mol), or a ratio of 1:550.000.

Many decades ago, this model was found to be able to predict the solubilities of both simple and more complicated organic molecules. In the most simple case, the solubility of heptane is lower than that of hexane due to the addition of one -CH_2- group. In the case of alkane molecules, a linear relation between the solubility and the number of -CH_2- groups is found (Birdi, 1997).

This model thus is based on the following assumptions when the alkane molecule is placed in water:

Alkane(CCC) is placed in a cavity in the water(ww).

wwwwwwwwwwwwwwwwwwwwwwwwwwwwwwwww
wwwwwwwwwwwwCCCCCCCCCwwwwwwwwwwwww
wwwwwwwwwwwwwwwwwwwwwwwwwwwwwwwww

The energy needed to create a surface area of the cavity will be proportional to the degree of solubility of the alkane. Thus, the solubility of any alkane molecule will be given as

$$\text{Free energy of solubility} = \text{Proportional to the cavity surface area} \quad \text{(surface tension of the cavity)} \quad \text{(B.1)}$$

By analyzing the solubility data of a whole range of alkane molecules, the following relation was found to fit the experimental data:

$$\text{Free energy of solubility} = \Delta G^\circ_{sol} = R\,T\,\log\,(\text{solubility}) \qquad \text{(B.2)}$$

$$= (\gamma_{cavity})\,(\text{S area alkane}) \qquad \text{(B.3)}$$

$$= 25.5\,(\text{S area alkane}) \qquad \text{(B.4)}$$

For solubility of alkanes in water, the total surface area, TSA, gives the solubility:

$$\ln\,(\text{sol}) = -0.043\,(\text{TSA}) + 11.78 \qquad \text{(B.5)}$$

where solubility (sol) is in molar units and TSA in Å^2. For example,

Alkane	(sol)	TSA	Predicted (sol)
N-Butane	0.00234	255	0.00143
n-Pentane	0.00054	287	0.0004
n-Hexane	0.0001	310	0.0001
n-Butanol	1.0	272	0.82
N-Pentanol	0.26	304	0.21
N-Hexanol	0.06	336	0.05

Note: The constant 0.043 is equal to γ cavity/RT = 25.5/600.

The surface areas of each group in n-nonanol ($C_9H_{19}OH$) were estimated by different methods. These data are as follows:

CH$_3$	CH$_2$	CH$_2$	CH$_2$	CH$_2$	CH$_2$	CH$_2$	CH$_2$	CH$_2$	OH
85	43	32	32	32	32	32	40	45	59

It is seen that, if one needs the TSA of n-decanol, then it will be [TSA of nonanol + TSA of CH$_2$] 431 + 32 = 463 Å^2.

The data for solubility of homologous series of n-alcohols is of interest, as shown in the following table.

Alcohol	Solubility (mol/L)	Log (S)
C$_4$OH	0.97	0.013
C$_5$OH	0.25	-0.60
C$_6$OH	0.06	-1.22
C$_7$OH	0.015	-1.83
C$_8$OH	0.004	-2.42
C$_9$OH	0.001	-3.01
C$_{10}$OH	0.00023	-3.63

Accordingly, this algorithm allows one to estimate the solubility of water of an organic substance. The estimated solubility of cholesterol was almost in accord with the experimental data (Birdi, 2008).

It is seen that log (S) is a linear function of number of the carbon atoms in the alcohol. Each -CH$_2$- group reduces log (S) with a 0.06 unit.

Appendix C: Gas Adsorption on Solid Surface—Theory

Gas molecules will adsorb and desorb at the solid surface. At equilibrium, the rates of adsorption (R_{ads}) and desorption (R_{des}) will be equal. The surface can be described as consisting of

Total surface area $= A_t = A_o + A_m$
Area of clean surface $= A_o$
Area covered with gas $= A_m$
Enthalpy of adsorption $= E_{ads}$

The following relations can be written:

$$R_{ads} = ka\, p\, A_o \qquad (C.1)$$

$$R_{des} = kb\, A_m \exp(-E_{ads}/RT) \qquad (C.2)$$

where ka and kb are constants.
  At equilibrium,

$$R_{ads} = R_{des} \qquad (C.3)$$

and the magnitude of A_o is a constant.
  Further we have the following:

Amount of gas adsorbed $= N_s$
Monolayer capacity of the solid surface $= N_{sm}$

By combining these relations

$$N_s/N_{sm} = A_m = A_t \qquad (C.4)$$

we get the well-known Langmuir adsorption equation (Birdi, 2008):

$$N_s = N_{sm}/(a\, p)/(1 + a\, p)) \qquad (C.5)$$

Additionally, the heat of adsorption has been investigated. For example, the amount of Kr adsorbed on AgI increases when temperature is decreased from 79 K (0.13 cc/g) to 77 K (0.16 cc/g). These data allow one to estimate the isosteric heat of adsorption (Jaycock and Parfitt, 1981):

$$(d\ (Ln\ P/dT) = q_{ads}/RT^2 \qquad (C.6)$$

The magnitude of qads was in the range of 10 to 20 kJ/mol.

Appendix D: Common Fundamental Constants

Angstrom, $\text{Å} = 10^{-8}$ cm $= 10^{-10}$ m

Micrometer, μm $= 10^{-6}$ m

Boltzmann constant, $k_B = 1.381 \times 10^{-23}$ J K^{-1}

Electronic charge, e $= 1.602 \times 10^{-19}$ C

Ideal gas law: P V $=$ k T $= 4.12 \times 10^{-22}$1 J at 298 K (25°C)

1 cal $= 4.184$ J

1 erg $= 10^{-7}$ J

1 atm $= 1.013 \times 10^5$ N m^{-2} (Pa)

kT/e $= 25.7$ mV (at 25°C)

Permittivity of free space, eo $= 8.854 \times 10^{-12}$ C^2 J^{-1} m^{-1}

eV $= 1.6021 \; 10^{-19}$ J

Kilowatt hour $=$ kWh $= 3.6 \; 10^6$ J

Viscosity of water $= 0.001$ N s m^{-2} $= 1$ at 20°C

Dielectric constant of water $= 80.2$ at 20°C

Index

Milton Keynes UK
Ingram Content Group UK Ltd.
UKHW040106071024
449327UK00019B/853